Complete VCR Troubleshooting and Repair

Written by:
Joseph Desposito
and
Kevin Garabedian

A Division of Howard W. Sams & Company
A Bell Atlantic Company
Indianapolis, IN

PROMPT© Publications is an imprint of Howard W. Sams & Company, A Bell Atlantic Company, 2647 Waterfront Parkway, E. Dr., Indianapolis, IN 46214-2041.

International Standard Book Number: 0-7906-1102-3

Acquisitions Editor: Candace M. Hall

Editor: Natalie F. Harris

Assistant Editors: Pat Brady, Loretta L. Leisure

Typesetting: Leah Marckel

Indexing: Loretta L. Leisure

Cover Design: Christy Pierce

Graphics Conversion: Terry Varvel

Additional Illustrations: Courtesy of Thomson Consumer Electronics

PRINTED IN THE UNITED STATES OF AMERICA

9 8 7 6 5 4 3 2 1

Contents

Dedication

This book is dedicated to Lorraine Desposito for the extraordinary help she provided during this project.

Preface

r so, VCRs have worked themselves into the majority of U.S. ıomes have more than one. What to do when a VCR goes on ːmium price to repair it? Junk it for a new model at a cost of ɛ. But, if you are like we are, you would like to be able to fix it. ricans tend to be a throw-away society, there is also a desire in find out what's wrong and fix it, if possible. That's where this

lie to say that VCR repair is easy. It's not. But, it is not so compli- ː beyond the ability of anyone who makes an honest effort to learn ʋork and what can go wrong. Add to this mix the proper tools, and you ı chance at spending a few dollars to repair a problem, rather than $50 more. This book teaches you to find and repair problems in VHS VCRs, ːommon type of VCRs in use today. Many of the principles presented ..ere are useful , too, for repairing other kinds of VCRs (8 mm VCRs and older Beta models) and other electronics products in general. After all, if you learn how to fix a power supply problem in a VCR, you probably will be able to repair similar problems in TVs, fax machines and other electronics.

This book uses a hands-on approach to help you understand what is involved in VCR repair. You won't see "press release" pictures of tools and test equipment. You'll see action pictures showing you how a specific tool or piece of test equipment is used to solve a problem.

We also try to bridge the gap between VCR troubleshooting theory and practice. To solve a problem with any of the circuits in a VCR, you have to know, in general, how they operate. To this end, we supply sample schematics and explain in a high-level way how the circuit works. We don't bog you down with every engineering detail of a circuit.

Additionally, we tell you from our own experience how likely it is that a particular circuit may fail. This technique will help you get right to the heart of a failure, rather than searching all over the VCR in places where failures are less likely to occur.

For every schematic, we also provide a photo of a sample VCR circuit to show you how the schematic representation of the circuit looks in a real VCR. This will help you find parts like the tuner or head amplifier when you work on your own VCR.

We know from experience that most VCR repairs can be done without expensive equipment. If you are just getting started, you probably do not want to invest a whole lot of money in sophisticated analyzers. If you are wondering if you can repair a VCR without an oscilloscope at your side, for example, rest assured that you can. At the same time, we recognize the fact that you may have access to expensive equipment, even if you do not own it. Therefore, you will find some instruction in the book about how to use high-end equipment.

In the last half of the book we present real-life repair examples on different VCR models. A close reading of these case studies will impress upon you the variation to be found when repairing VCRs. The same problem that is relatively easy to fix in one VCR might be a nightmare in another. It really depends on the problem and how the VCR is constructed. We hope to provide you with a good feel for solving all kinds of VCR problems rather than giving you cookie-cutter solutions.

VCR repair can be a fun and interesting sideline as well as the basis of a life-long profession. The VCR itself is quite an extraordinary piece of electronics equipment encompassing the latest in electronics and mechanical techniques. Learning to repair a VCR will also help you gain an understanding of how modern electronic circuits work together with mechanical assemblies to produce a sophisticated piece of consumer electronics equipment.

A book such as this one cannot be produced by one or two people. In this regard, we want to sincerely thank Candace Hall, Natalie Harris and all the people at PROMPT® Publications who helped to bring this book to market.

Chapter 1

Overview of VCR Circuits

A video cassette recorder, or VCR as everyone calls it, is a wonderful combination of electronic and mechanical wizardry. The VCR circuits, which are the subject of this chapter, perform a whole host of tasks, for example, processing video and audio signals, controlling motors, and supplying power to the machine.

It's natural to think of a VCR as a machine that can do two things exceedingly well: 1) Play a video tape and 2) record a video signal onto a video tape. **Figures 1.1** and **Figure 1.2** are simplified diagrams showing the circuits and components of the VCR involved in these two operations. In each of the figures, inputs on the left and outputs on the right provide a better idea of the flow of the operation. As we go through this chapter, we will expand on these diagrams to help you gain a better understanding of how each section of a VCR works.

We begin our overview of VCR electronic circuits with the power supply. The power supply gives life to the rest of the circuits in the VCR and so we place it first in order.

1.1 Power Supply Circuits

As its name suggests, the role of the power supply is to provide power to the rest of the circuits in the VCR. The power cord of the power supply plugs into an AC (alternating current) outlet. In the United States, the AC line provides approximately 120 volts (V), alternating at a rate of 60 cycles per second (Hz), expressed 120 VAC. The job of the power supply circuitry is to transform this 120 VAC into the various AC and DC (direct current) voltages required by the other circuits in the VCR.

There are two basic types of power supplies: linear and switch mode. In a linear

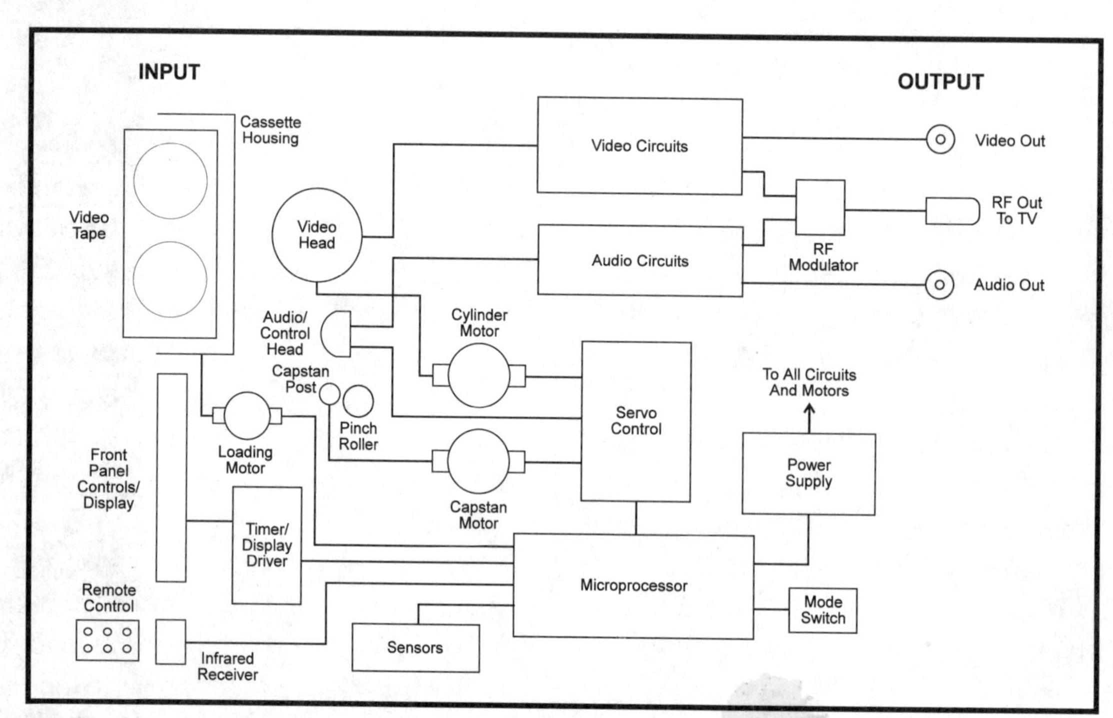

Figure 1-1. A simplified block diagram showing how a VCR plays a video tape.

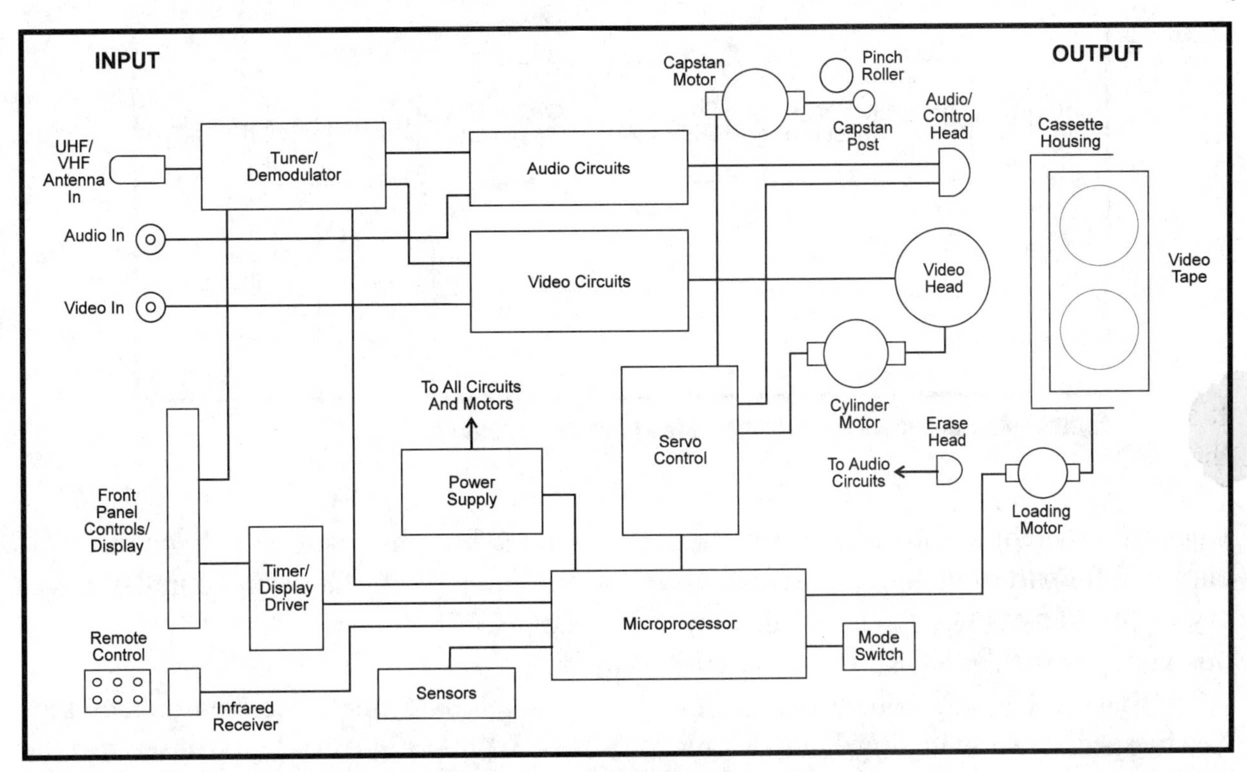

Figure 1-2. A simplified block diagram showing how a VCR records on a video tape.

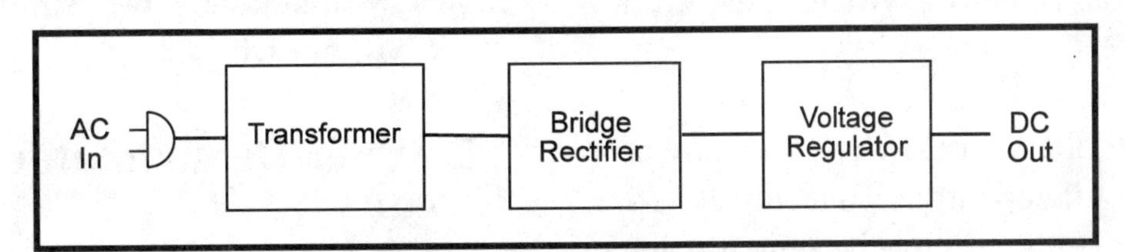

Figure 1-3. A simplified block diagram of a linear power supply.

supply, the AC line voltage is fed to the primary of a transformer. The secondary of the transformer produces the various voltages required by the VCR. Full-wave (bridge) and half-wave rectifiers change the voltages from AC to DC. These "raw" DC voltages go to voltage regulators, which produce regulated DC voltages. Linear supplies have been around longer than switch mode supplies. You'll find them in older VCRs as well as in some newer models. **Figure 1.3** shows a simplified block diagram of a linear power supply.

The newer switch mode type of power supply works like a switch. Voltage from the AC line passes through an rf filter, which is a small coil with capacitors connected to a ground. The purpose of the filter is to suppress electromagnetic interference (EMI). The AC voltage is rectified by a bridge rectifier and increased to produce a high DC voltage. A switching transistor converts this DC voltage back to a square wave type AC voltage. The switching frequency of the transistor varies from about 50 to 100 kHz, depending on the load. This switching voltage is fed to a switch mode transformer.

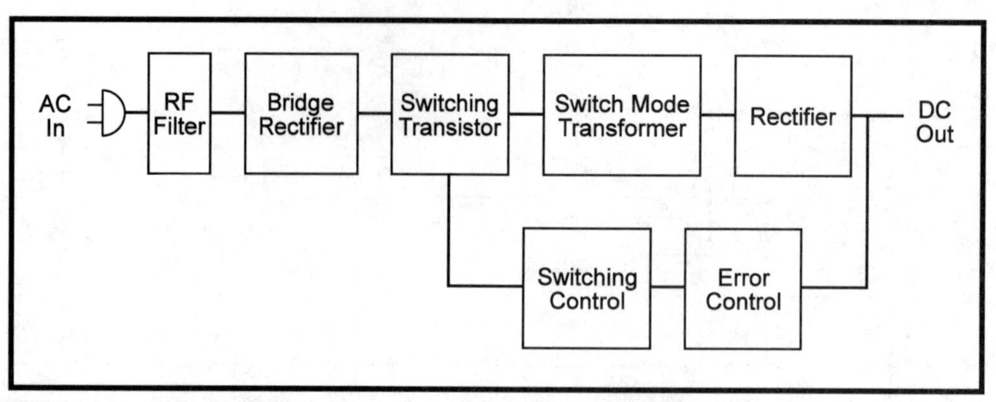

Figure 1-4. A simplified block diagram of a switch mode power supply.

Due to the switching, the transformer takes current only part of the time. The secondary of the transformer produces the various voltages for the VCR. These are still AC voltages. The AC voltages are rectified by special fast switching diodes to create DC voltages again, and then strictly regulated to produce the voltages required by the VCR.

The output voltage in a switch mode power supply is regulated through a feedback circuit. This feedback circuit typically includes an optoisolator that isolates the secondary side from the primary side of the power supply. A simplified diagram of a switch mode power supply is shown in **Figure 1.4**.

You may think that the power supply turns on only when you press the power supply switch. But this is not the case. A part of the power supply must be independent of the power switch so that it can stay on at all times. The VCR's clock, for example, needs power even when the VCR is off.

Keep in mind that the power supply always contains a fuse to disconnect power from the AC line if necessary. A blown fuse usually indicates further trouble in the power supply.

If you look again at the diagrams of **Figure 1.1** and **Figure 1.2**, you'll see that the power supply is shown in both. This is an indication that if the power supply dies, the VCR will not work at all.

1.2 Tuner/Demodulator Circuits

The tuner is a separate unit in a VCR. It is housed in a metal case that acts as a shield against interfering signals. When a channel is selected, the tuner mixes the signal from the antenna with the local oscillator and produces an intermediate frequency or IF signal (this is the superheterodyne principle). Since there are both audio and video signals involved, two IF signals are produced.

These IF signals are amplified and mixed to produce a 45.75 MHz video IF signal and a 4.5 MHz audio IF signal. The video IF signal is fed to the video detector and demodulated. This produces a composite video signal, which appears at the video

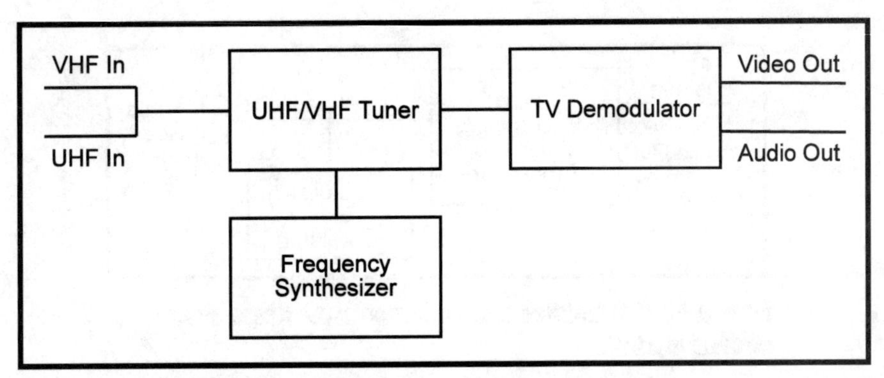

Figure 1-5. A simplified block diagram of the tuner/demodulator section of a VCR.

output jack (at the rear of most VCRs) and is also fed to the video circuitry for further processing. The audio IF signal is amplified and fed to the audio detector and demodulated. This audio appears at the audio output jacks of the VCR and is also sent to the audio circuits for further processing.

There are two basic types of tuners. One is the old-style rotary mechanical tuner, the other is the varactor tuner. The rotary tuner has a round drum inside with thirteen mounted coils, one for each channel. Fine tuning is accomplished by a small screw inside the coil that slightly changes the frequency of the oscillator.

The varactor tuner takes its name from a device it employs, the varactor diode. These diodes have the ability to change their capacitance according to the applied voltage. By changing the capacitance of the diode, the tuner tunes to a different channel.

There are two kinds of varactor tuners: voltage synthesized and frequency synthesized. Voltage synthesized tuners use small wheels for adjustment and are found in older VCRs. All newer VCR models use frequency synthesized tuners, also known as phase-locked-loop (PLL) tuners. No adjustments are necessary. **Figure 1.5** shows a simplified block diagram of the tuner/ demodulator section of a VCR.

1.3 Audio Circuits

You will find two types of audio circuits in VCRs: linear audio and hi-fi audio. Prerecorded tapes offer both kinds of audio. Thus, it is the audio capability of the VCR that determines what you hear.

Linear audio can be mono or stereo; hi-fi is only stereo. Linear sound uses a narrow portion of the tape. In recording mode, the audio signal is recorded by mixing it with a signal from a bias generator. This generator produces a high-frequency voltage at about 50 kHz, four times the highest frequency of the audio signal. In the process of recording, part of this signal biases the tape, and part of it erases the old recording. The signal is then sent to the recording head (a universal head for playback and record). The bias voltage is adjusted for best recording, not too high, not too low. This keeps the tape in the center of the hysteresis curve to avoid distortion.

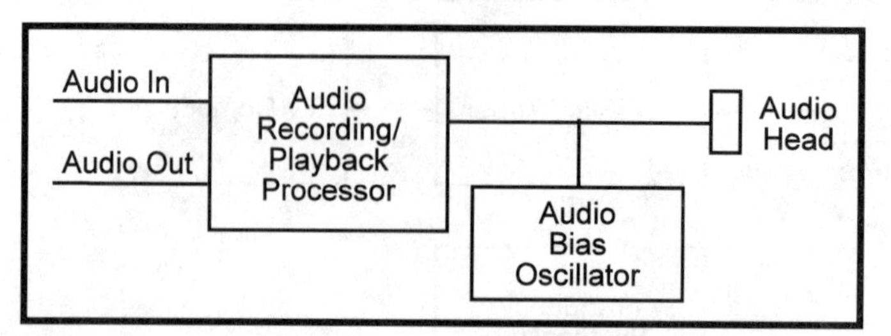

Figure 1-6. A simplified block diagram of the linear audio circuits of a VCR.

When the tape is actually recorded, it is magnetized by the audio signal and the bias voltage. But the bias signal is of much higher frequency. The playback head cannot pick up the high frequency bias signal and plays only the audio signal. This signal, which has a very low level, travels to the head amplifier. Then it is amplified again and goes to the rf converter, where it is converted to channel 3 or 4.

Audio also comes into the VCR from the antenna through the tuner. The tuner produces an IF signal, which goes to the IF amplifier. The sound is split from the IF signal and goes to the sound IF and to the sound detector. The sound detector produces the audio signal, which is amplified and sent to the rf converter through a switch. The switch determines the mode of the VCR: playback or tuner. **Figure 1.6** shows a simplified block diagram of the linear audio circuits.

The hi-fi circuitry is quite different. Before recording, the signal is frequency modulated. This frequency modulated signal is applied to two heads mounted on the upper part of the video drum. The heads spin in the opposite direction to the movement of the tape. The resulting speed is the sum of the two speeds, approximately one yard per second. The video head makes 1,800 revolutions per minute. The higher speed improves the signal.

On the playback side, the frequency modulated (FM) signal is picked up by the two audio heads and goes to the head amplifier. The signal gets amplified and then goes to the detector, where the FM signal is demodulated to obtain the audio. This complicated process gives the video recorder a high quality sound.

1.4 Video Circuits

When you record onto a video tape, the VCR takes the video signal from the tuner/demodulator or VIDEO-IN jack and sends it to a chroma/luma processor. This processor separates the video signal into luminance (brightness) and chroma (color) signals. The luminance signal is frequency modulated, while the chroma signal is downconverted from 3.58 MHz to 629 kHz. This needs to be done in order to "fit" the video information on the tape. The two signals are then amplified and sent to the video heads to be recorded on the video tape.

When you play a tape, the video signal is picked up from the tape by the video heads. This signal has a very low level, in the mil-

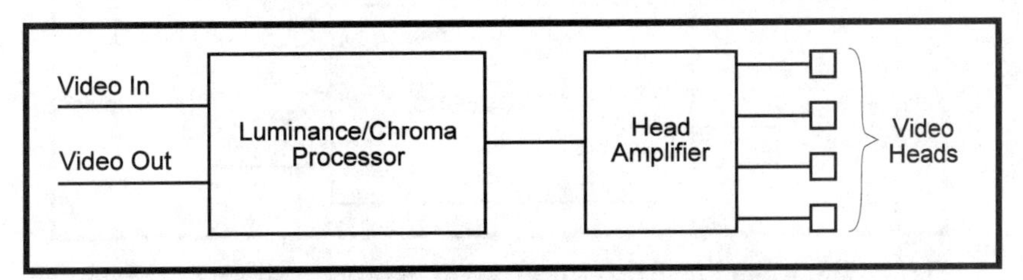

Figure 1-7. A simplified block diagram of the video circuits of a VCR.

livolt range. From the video heads, the signal goes to the video head amplifier. Each video head signal (typically two or four heads) goes to a separate input and the signals are mixed inside the amplifier.

From there, the signal is sent to the chroma/luma processor. This circuit separates the signal into the FM luminance signal and downconverted chroma signal. Then it demodulates the luminance signal and restores the chroma signal to its correct frequency. The luminance and chroma are then mixed to form the video signal.

The video signal goes to the video output terminal of the VCR and also to the rf modulator. The signal is modulated, which makes it accessible to televisions which do not have direct video inputs. This output is switchable to channel 3 or 4. **Figure 1.7** shows a simplified block diagram of the video circuits.

1.5 Microprocessor

A microprocessor is a circuit that controls the operation of a VCR. A microprocessor has inputs which receive signals from various sensors and switches, and outputs which exercise control over the various circuits of the VCR.

A microprocessor typically has 60 or more pins. Signals on these pins may be high (+5 V), low (0 V), pulse, or a variable analog voltage. Analog signals are sent to an analog-to-digital (A/D) converter within the microprocessor for digitizing. Most pins are classified as input (I) or output (O), though some may be bi-directional (I/O).

Microprocessor signals can be active high or active low. An active high signal becomes active when the voltage at the pin is +5 V. An active low signal becomes active when the voltage at the pin is at 0 V. An active low signal is denoted by a bar over the name of the signal, for example RESET.

Microprocessors are customized for specific VCRs. If you ever have to replace a microprocessor, you must do it with an exact replacement part; nothing else will work.

When troubleshooting a VCR, it may appear as though the microprocessor is at fault. For example, you press PLAY and the VCR rewinds the tape. The microprocessor is simply responding to an input signal in a programmed way. If the input signal is not correct, the output will be wrong, too. In a case like this, the problem is more

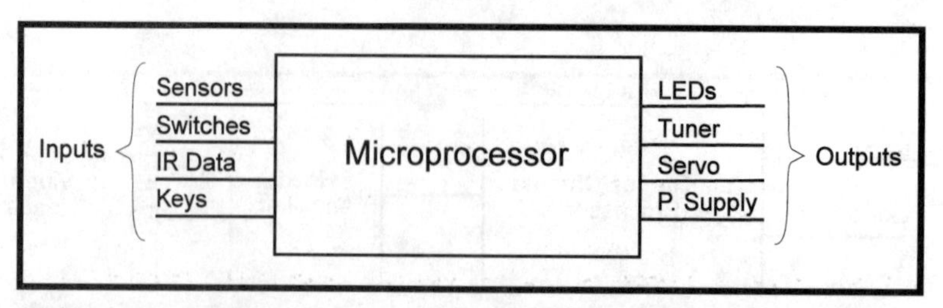

Figure 1-8. A simplified block diagram of the microprocessor showing several inputs and outputs.

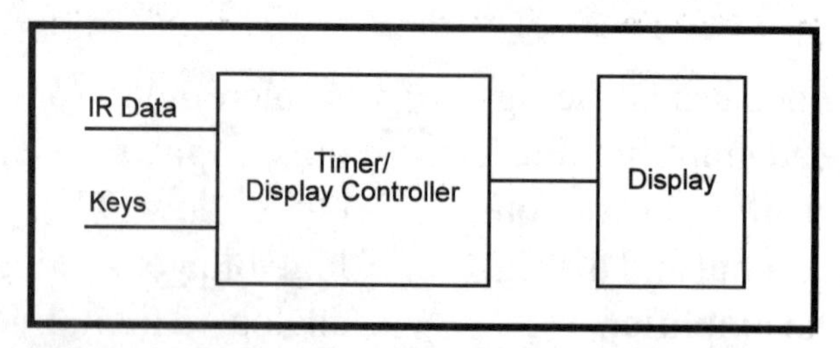

Figure 1-9. A simplified block diagram of the VCR display circuits.

likely to be caused by a faulty mode switch (which sends signals to the microprocessor) than by a faulty microprocessor. **Figure 1.8** shows a simplified block diagram of the microprocessor with typical inputs and outputs.

1.6 Display

VCRs have either a fluorescent light vacuum tube display or a liquid crystal display (LCD). The fluorescent type has a filament, which obtains power from the power supply. The LCD doesn't have a filament. In either case, the VCR display is connected to a timer/display controller. **Figure 1.9** is a simplified diagram of the display circuits.

The display is an excellent visual indication of whether or not the VCR is receiving power. If the VCR is plugged in, the display should light. If it doesn't, this usually indicates a problem with the power supply.

1.7 Remote Control Circuits

The remote control circuits consist of an infrared transmitter and infrared receiver. The infrared transmitter is located in the handheld remote control unit. When you press a button on this handheld unit, it generates an infrared signal that is transmitted to the infrared receiver. The receiver is located behind tinted glass on the front panel of the VCR.

The infrared signal that comes from the handheld unit is a serial signal coded by the manufacturer. Once it reaches the VCR, the coded signal is sent to a D/A converter. This converts the digital pulses to an analog signal, which is used to control the VCR.

Each button, or control, on the handheld unit has a different code. There are two types of controls: momentary and continuous. Momentary controls are used, for instance, to change channels, while continu-

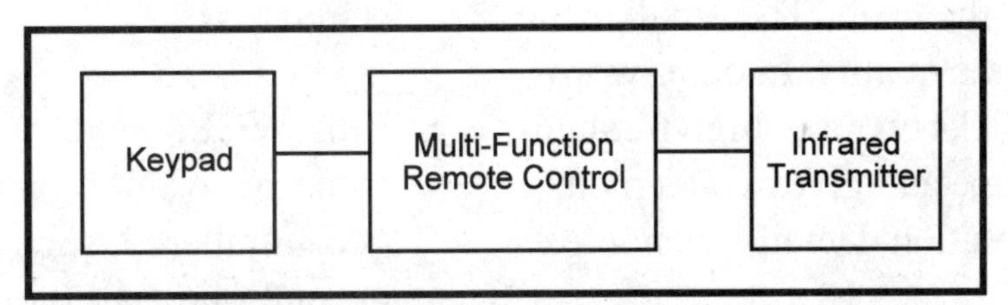

Figure 1-10. A simplified block diagram of the remote control circuits of a VCR.

ous controls, for example, are used to increase or decrease the volume. **Figure 1.10** is a simplified diagram of the remote control circuits.

Today's remote control units are elaborate and perform multiple tasks. Remote control handhelds are not interchangeable, though you can purchase limited-function programmable remotes.

1.8 VCR Servo System

VCRs contain two types of servos. One is called the capstan servo. It controls the speed of the tape. The other is called the drum servo. It controls the speed of the drum motor.

Two reference signals ensure that the capstan servo does its duty. The first signal, CTL, comes from the control head. The control head reads the control track, which is recorded on the bottom of the video tape. The second signal is taken from the capstan motor; it is called the frequency generator (FG) signal.

The capstan motor turns the capstan shaft. The capstan shaft works together with the pinch roller to control the speed of the tape coming out of the cassette.

The operation of the drum (cylinder) motor servo is also based upon two reference signals. The drum servo uses vertical sync and FG signals. During playback, vertical sync is taken from a clock generator, controlled by a crystal. An FG signal is produced by a magnetic pickup coil, which is mounted on the bottom of the drum motor.

On some newer model VCRs, the FG signal is produced by a circuit trace on the motor's printed circuit board (PCB). To track video heads precisely at the beginning of a video track, a pulse pickup is mounted on the motor's PCB. The pulse pickup is usually a Hall effect integrated circuit (IC).

When the servo controls are not functioning, most VCRs mute the sound. Servos are closed-loop circuits. Troubleshooting them is always a challenge. **Figure 1.11** is a simplified diagram of the capstan and drum motor servo system.

1.9 Recording and Playback Heads

VCRs have universal heads. Mechanically, it's not possible to separate recording heads and playback heads. The important quality of heads are what materials they are constructed of, how they are manufactured, and the gap in the head. The smaller the gap, the better the quality. In other words, a smaller gap picks up more high frequencies from the tape and gives you a picture with better resolution. Details are not lost.

Video heads and hi-fi audio heads are attached to the video drum. There are also separate audio and control heads. The control head keeps track of how fast the video drum is turning. It has to turn at a rate of 1,800 revolutions per minute (30 revolutions per second) to conform to industry standards.

Most VCRs have two or four heads, but some have three or five. When they are odd-numbered, such as three, two are for normal play and the third one is used for pause and slow motion. When there are five heads, four are used for normal playback, the fifth one for pause and slow motion. This technique provides exceptional quality. **Figure 1.12** is a close-up photo of a video head.

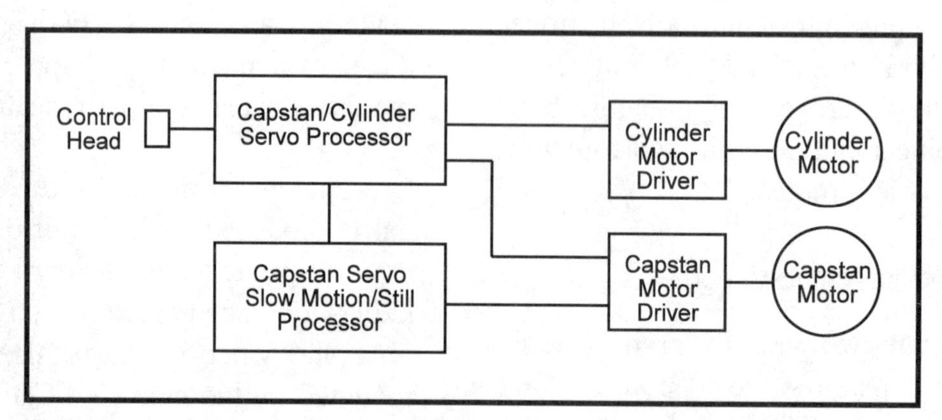

Figure 1-11. A simplified block diagram of the capstan servo system of a VCR.

Figure 1-12. A close-up photo of a video head.

Chapter 2

Tools and Test Equipment

Though VCRs are complex, you don't need complex tools or test equipment to repair them. You can perform most repairs with a small selection of gear, which will be described in this chapter. Of course, some repairs demand more sophisticated equipment, but these are the exception rather than the norm.

2.1 Tools

First and foremost, when repairing VCRs or any other kind of electronic equipment, you need a good soldering iron. An iron rated at 25 to 30 watts (W) with a pencil tip is perfect for the job. A Weller soldering iron is a good choice—you can find them in parts stores and advertised in electronics repair catalogs, but other brands will do. You should also purchase a stand for the iron. This will give you a handy spot to place your iron so that you don't accidentally burn any materials around your workbench.

A tool that goes hand in hand with a soldering iron is a desoldering tool. This vacuum pump type tool "sucks up" melted solder, thus allowing you to remove defective components from a printed circuit board. If a desoldering tool doesn't remove all the solder needed to free a pin or lead, you can remove any remaining solder with a solder wick or braid.

Good soldering and desoldering skills are an absolute necessity for performing VCR (or any other electronic equipment) repairs. It's not enough to be able to remove defective resistors and capacitors and solder in replacements. You need to be able to replace integrated circuits, some with 40 or

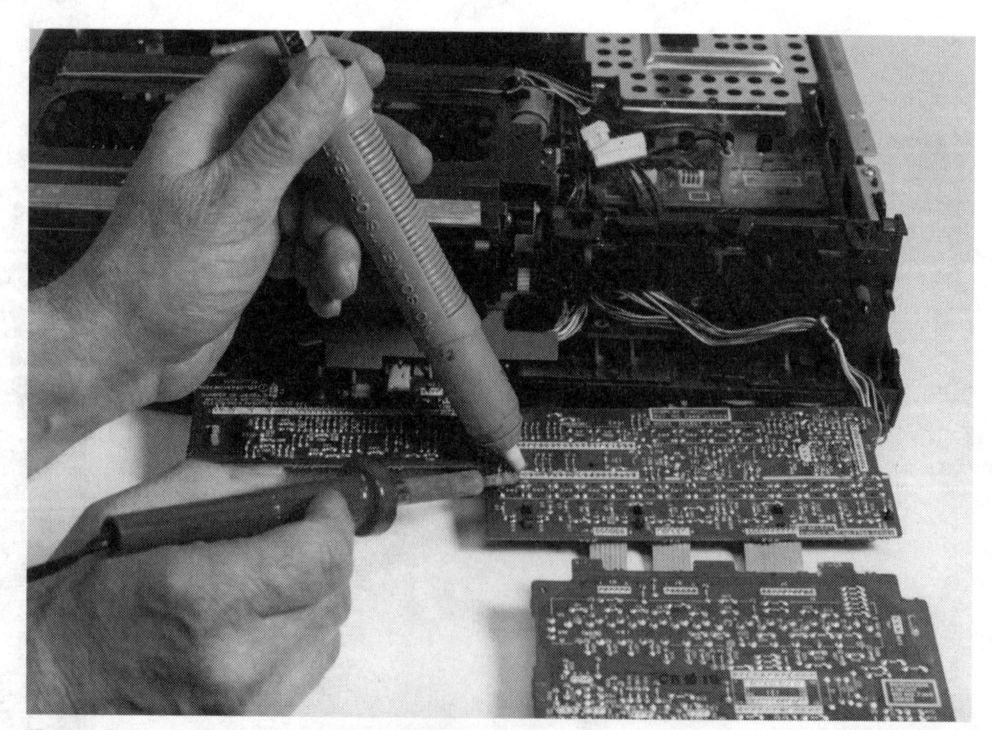

Figure 2-1. A VCR's timer IC being desoldered with a soldering iron and a desoldering tool.

more pins! **Figure 2.1** shows a timer IC being desoldered with a soldering iron and a desoldering tool.

You will need a set of good Phillips screwdrivers since most of the screws in the VCR have Phillips heads. Most VCR repairs start with the removal of the top cover of the VCR as shown in **Figure 2.2**. Many repairs also demand that you remove the bottom cover as shown in **Figure 2.3**. You'll need screwdrivers of various shaft lengths and with magnetized heads. Many repairs require the removal of the cassette housing. This is the part of the VCR where you place the cassette tape in the machine. To remove the cassette housing, you need to remove the screws that attach the housing to the VCR chassis. These screws are often in tight places. You need a long screwdriver shaft to get to them and a magnetized head to bring the screw up and out of the VCR. If you lose a screw in a VCR, you are asking for trouble. A dropped screw can easily become stuck among the many mechanical parts of the VCR. If you then try to play the VCR, you may jam parts, damage motors, and so forth. **Figure 2.4** shows how a Phillips head screwdriver is used to remove a screw from a cassette housing.

Figure 2-2. Removing the top cover of a VCR with a Philips head screwdriver.

A small flathead screwdriver is convenient to have. These are useful more for their functional flexibility than for the need to unscrew small flathead screws. This type of screwdriver is great for prying out ICs, as shown in **Figure 2.5**. An IC puller is not really needed. This tool is good when you have socketed ICs, but most VCR ICs are

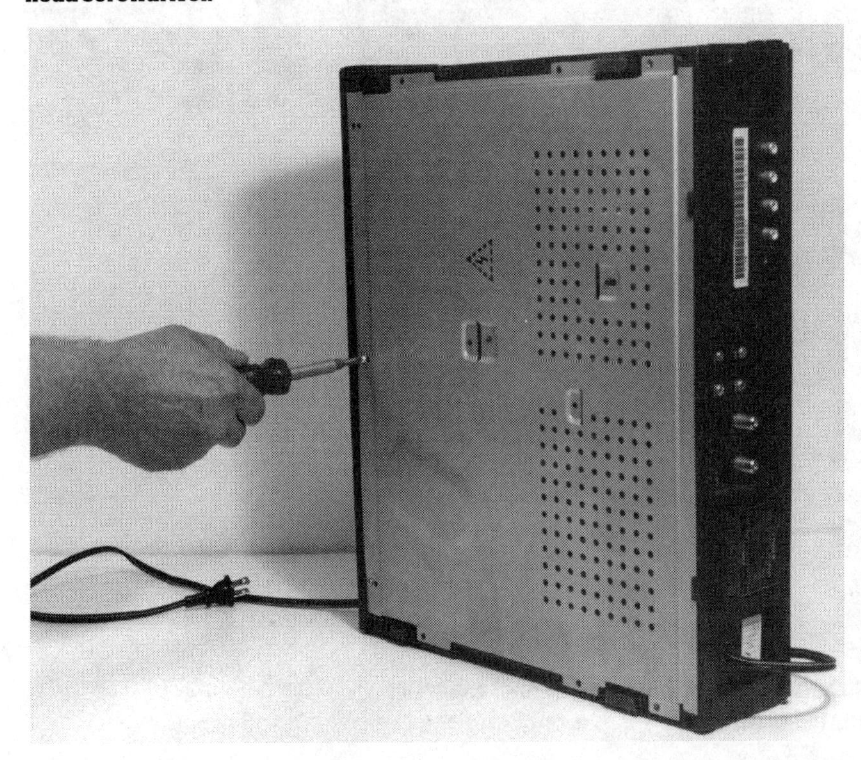

Figure 2-3. The bottom cover of a VCR can contain as many as ten Philips head screws.

Figure 2-4. A magnetized Philips head screwdriver is essential for tight spots such as removing the cassette housing.

Figure 2-5. A small flathead screwdriver can be used to pry out ICs.

Figure 2-6. A long-nosed pliers is the perfect tool for removing a small capacitor from the power supply after it has been desoldered.

Figure 2-7. The special alignment tool shown here is being used to adjust the tape post guide; this is done in conjunction with an oscilloscope.

soldered into a PC board. Once you desolder the IC, it needs just a little bit of force to remove it (if you have to exert a lot of force, you'd better do some more desoldering). A small flathead screwdriver is perfect for this job.

Before we move on to other tools, you may want to consider one more type of screwdriver—a cordless power model. If you intend to repair a great number of VCRs, you may welcome the added convenience a cordless power screwdriver provides. After all, some VCR bottom covers are loaded with screws.

Long-nosed pliers are needed to release springs, belts, and so forth. They are also good for removing components that have just been desoldered and are hard to reach. **Figure 2.6** shows the removal of a small electrolytic capacitor from a switch mode power supply. The pliers are also needed to place the new capacitor into the PC board. Tweezers are useful for the same kind of jobs as long-nosed pliers, but can reach into tighter spots.

A set of small-size hex wrenches are useful for video alignments and brake adjustments. Also, there are special tools for adjusting the audio head and the tape post guides. **Figure 2.7** shows one of these tools in action, adjusting the tape post guide. This is done in conjunction with an oscilloscope. As you turn the tape post, the waveform on the scope changes. We'll tell you more about this alignment procedure later in the book.

A good quality test tape is useful for aligning the video heads and audio heads, as well as the video circuits of the VCR. The test tape lets you test for speed, wow, and flutter. This is done in conjunction with a frequency counter. You hook up a fre-

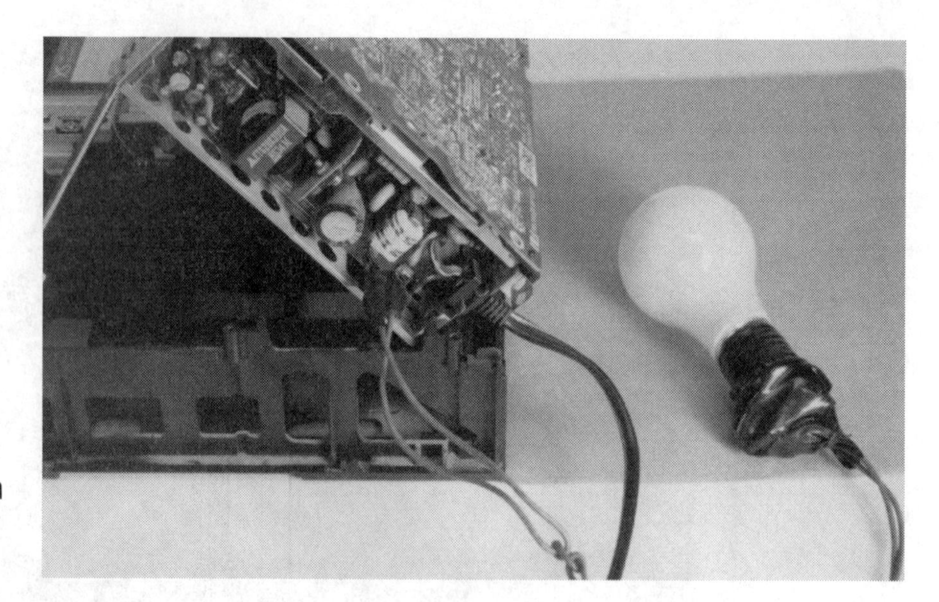

Figure 2-8. A socketed light bulb can be used in place of a fuse when testing a switch mode power supply.

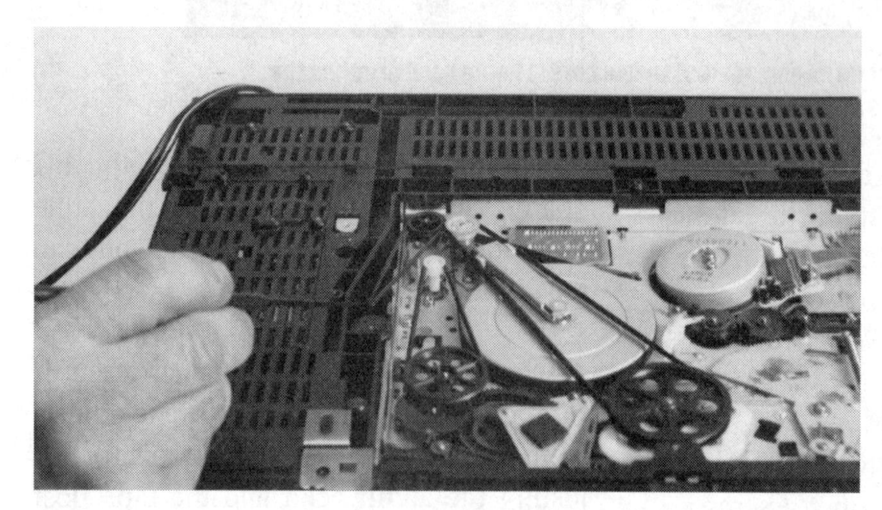

Figure 2-9. A special hook tool is handy for working with springs and belts.

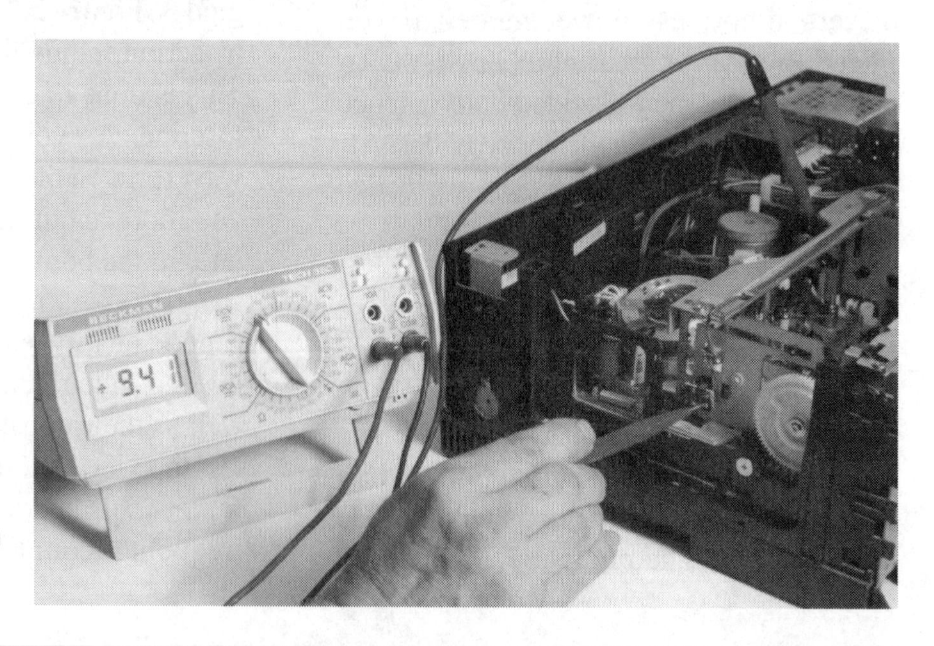

Figure 2-10. A bench type DMM measuring a voltage in a switch mode power supply.

quency counter to the audio output and then read the frequency. If the frequency keeps changing, that means the transport has a problem.

A good tool for servicing a VCR is one that you may not even consider as a tool. It's a socketed light bulb (about 90 W) with alligator clips on the leads coming out of the socket (**Figure 2.8**). This is used as a replacement for a blown fuse when testing power supplies. If a power supply blows a fuse, you may be tempted to put in a replacement fuse. If you do this without figuring out the problem in the supply, it's very likely that you will blow the replacement fuse, too. Clipping the leads of the light bulb onto each end of the fuse holder allows you to test for a short without blowing a fuse. Instead, the light bulb glows brightly.

A specialized tool for working with small belts and springs is shown in **Figure 2.9**. This tool is handy and inexpensive, but long-nosed pliers will serve you just as well. Another specialized tool is a pair of pliers that have an action opposite to that of regular pliers. In other words, when you squeeze the pliers, they open instead of close. These pliers are useful for releasing spring washers. For example, when you remove the clutch assembly in a VCR, you have to remove clips, so you need these special pliers.

A tool that makes it easy to observe the mechanical workings of a VCR is a test jig. This jig looks like a transparent video cassette without the tape. You place this jig into the VCR essentially to "fool" the electronics into responding as though a real tape had been placed in the machine. Once the jig goes in, you can operate the VCR controls such as play, record, fast forward, and rewind. Some jigs will not work in some VCRs due to the positions of the sensors. If a jig won't load into a VCR you can sometimes circumvent the problem by covering one or more sensors with your fingers.

Other tools you should have on hand are wire cutters and strippers, an X-Acto type knife, and a small flashlight. Exotic tools such as a tension meter may be nice to have, but are expensive. Amateurs are not likely to own this tool, nor will they need it very much. Even professionals can get by most of the time without this type of tool.

2.2 Test Equipment

You'll be happy to know that the most important piece of test equipment by far for servicing VCRs is the ever popular digital multimeter, or DMM. If you are serious about servicing VCRs (if you are reading this book, you must be) take our advice and purchase a quality DMM, one that has features such as a diode test mode. A DMM is most useful for measuring DC and AC voltage and for measuring resistance. A DMM with the diode function can measure diodes and transistors for opens and shorts. **Figure 2.10** shows a bench type DMM at work measuring a voltage at one of the VCR's sensors. Handheld DMMs, such as the one shown in **Figure 2.11**, work just as well.

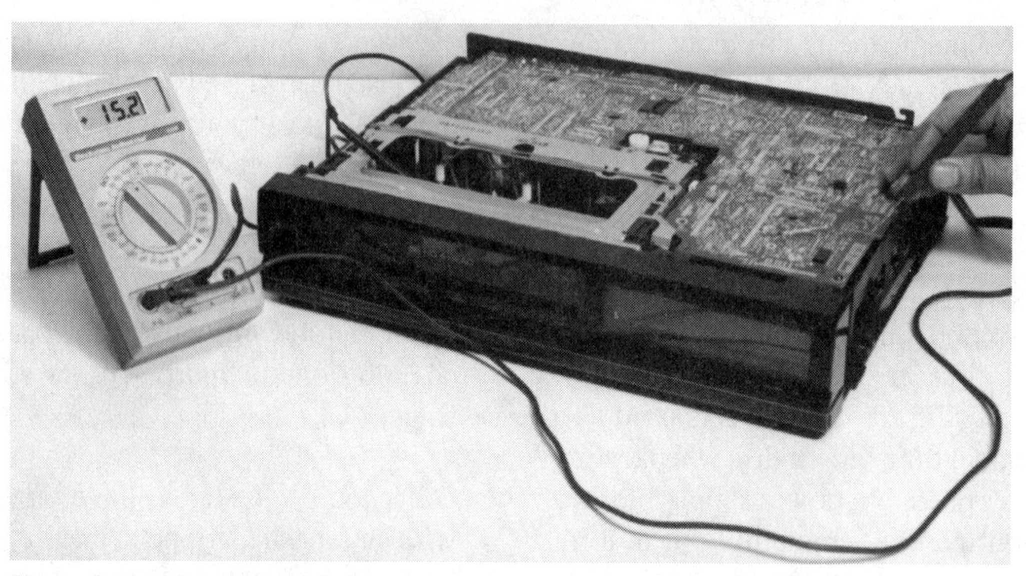

Figure 2-11. A handheld DMM measuring a voltage in a switch mode power supply.

A digital multimeter with a 200 VDC range is sufficient for testing VCRs. If you will also be using the DMM to troubleshoot televisions, you should get one that has up to a 1,500 VDC range. An analog meter works okay for certain jobs, but if it has a very low input resistance, it can change the characteristics of the circuit you are measuring. How do you tell? Either from the meter's manual or by placing the probes in a very sensitive area of the VCR. If the VCR stops working, the input resistance of the analog meter is low.

You may not think of the next piece of equipment as "test" equipment, but it is an absolute necessity for your test bench. This is a video monitor. When you are working with VCRs, you are always checking the video and audio outputs. You place a tape in the VCR, you look at the monitor. You connect a cable antenna to the VCR, you look at the monitor. A television serves the same purpose, but a video monitor does an all around better job in a test environment. Why? A video monitor connects to the VIDEO-OUT jack of the VCR. If you are experiencing a problem with the VCR's video, you can remove the RCA plug from the VIDEO-OUT jack and use it as a probe to check various test points on the printed circuit board where you expect to find a video signal. This is almost as good as using an oscilloscope for this job. Whereas the scope shows you the video waveform, the monitor shows you the video picture. You can't do this with a TV, though, because a TV can display only the modulated video coming from the VCR's rf modulator.

An oscilloscope is a glamorous tool, relatively expensive if you purchase a good one, and not even necessary for most VCR repairs. An oscilloscope lets you measure waveforms and is the perfect piece of equipment for making electrical adjustments. But, most VCRs don't really need to be "tuned up." If you plan to purchase an oscilloscope for VCR repairs, make sure the bandwidth is at least 60 MHz.

For safety purposes, you may want to invest in an isolation transformer. This piece of equipment sits between the VCR and the AC line, effectively isolating your body from the AC lines. This avoids potentially lethal shocks. If you are not knowledgeable at all about the safety precautions you should observe when working with electronic equipment, purchase the service manual for the VCR you intend to work on. All service manuals, as well as VCR*facts*® from Howard W. Sams & Co., explain safety procedures in detail.

Other kinds of test equipment, such as a power supply, transistor tester, logic probe, and frequency counter make for a well rounded test bench. You should also have on hand a supply of probes, test clips, test wires with alligator clips on one or both ends, and so forth. These test equipment accessories make it easier to do repairs.

2.3 Specialized Test Equipment

Specialized test equipment such as a Sencore video analyzer is very expensive, costing many thousands of dollars. This type of equipment is only for professionals, offering them an all-in-one piece of equipment, with special features for servicing VCRs. We use a Sencore universal video analyzer Model VA62A.

The VA62A is also useful for injecting signals into the VCR. For example, you can inject an rf signal into the tuner to check it. The VA62A has everything needed to align the IF frequency and to align the tuner. You can inject any channel. Also you can inject a test pattern, such as a crosshatch or a black and white pattern, to see how the video amplifier is working. Another feature of the VA62A is an audio generator with four different frequencies. This lets you add audio to the video signal, if required for testing. The VA62A also has a substitute for a servo signal, called the servo drive test.

We also use the VA62A when we suspect the video head is defective. The VA62A has an attachment called the VC63 VCR test accessory which tells you if the video head is good or no good. It tests the head dynamically when the VCR is working. It lets you know if the video head is the problem or the head amplifier or something else, such as the head switching circuits.

Chapter 3

General VCR Maintenance Procedures

A VCR, just like any other mechanical assembly, requires maintenance from time to time. Cleaning and lubrication should be performed at regular intervals, though most people run their VCRs until they "drop." A VCR may operate for several years before it experiences a problem. When a customer brings an older VCR into the shop for repair, you can be certain that the VCR also needs some general maintenance. If you fix the problem and don't attend to the maintenance, you may see the VCR back in the shop in a very short time. Worse yet, you may incur the wrath of the customer. If you are servicing your own or a friend's VCR, you don't want to have to fix it again right away. Not only is this a nuisance, but it also reflects on your ability (or inability) to do a thorough job.

3.1 Cleaning and Lubricating VCR Mechanisms

Whenever you play a tape, a little bit of dust is released. This is especially true of new tapes. The dust gets into the bearings of the capstan motor and on the surface of the rubber pinch roller. Eventually, these pieces get dirty or worn out. When a machine comes in for servicing, you have to pay attention to whether a part requires replacement or just needs cleaning. Generally speaking, the capstan shaft should be cleaned until it shines. You should not see any tape residue. The pinch roller should be cleaned as well as all the tape post guides. Again, these should be shiny; you should not see any trace of dust from the tape. Some VCRs have small posts that control the tape coming from the pinch roller and the capstan. These have to be cleaned, too.

After the cleaning is done, you should place a couple of drops of oil on the bearing of the capstan and the pinch roller and a drop each on the tape spindles. Don't stick an oil can into the VCR to do this lubrication. Instead, do what the old watchmakers used to do. They lubricated watches by dipping a horse hair (wedged into a match stick) into oil and then letting the oil run down the hair into the mechanism. You don't have to use a horse hair. You can use a tiny screwdriver instead. Just dip the screwdriver into oil and let it run down the shaft to the spot you want to lubricate. **Figure 3.1** shows how to lubricate the capstan bearing. Note that before you dip the screwdriver into the oil you should use it to lift the plastic washer at the base of the capstan shaft. Add the oil under this washer, not on top of it. The oil will run down into the capstan motor. A drop or two is all that is needed. Getting back to

Figure 3-1. Lubricating the capstan bearing by letting oil run down the shaft of a small screwdriver.

the spindles, make sure they turn easily and are not stuck. Sometimes oil or grease can become as sticky as glue over time. This makes the movement of the spindles very sluggish and adds additional tension to the mechanism. Special oil for VCRs can be purchased in any electronics supply store.

Whenever you press PLAY or RECORD, the tape guides in the VCR slide up like little robotic fingers playing cats cradle, pulling the tape up around the video head. The guides slide through a curved cutout in the chassis, supported by the metal on either side of the opening. This is a prime area for lubrication, not with oil, but with grease. Follow the old adage, "Oil anything that turns, grease anything that slides." The easiest way to do this is with a grease injector as shown in **Figure 3.2**. You should grease the right and left sides of both tracks on both sides of the chassis.

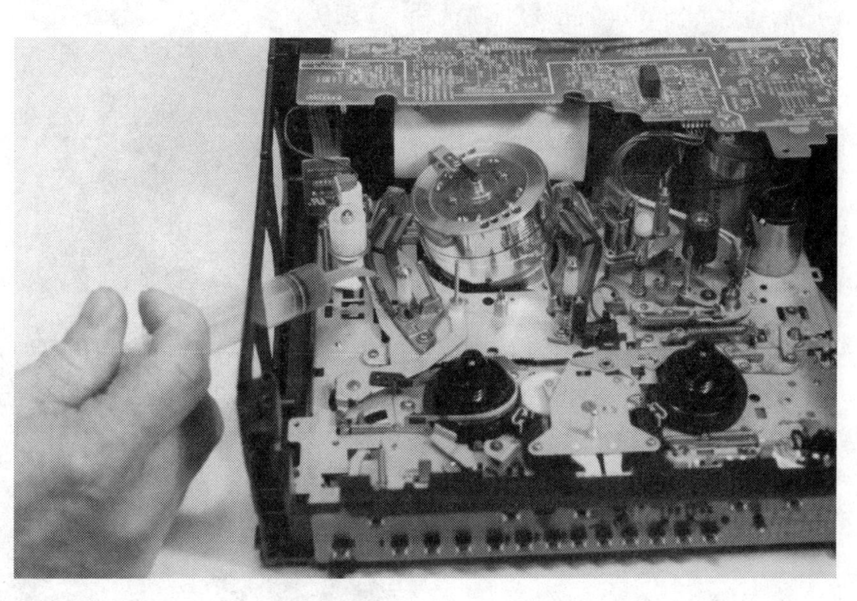

Figure 3-2. Greasing the tape guide tracks with a grease injector.

Cotton or foam swabs can be used for general cleaning but never to clean a video head. For example, a cotton swab dipped into alcohol is perfect for cleaning belts and tires.

The older the machine, the more likely you are to see cobwebs and large buildups of dust in the nooks and crannies of the VCR. If a VCR has been left in the closet for a while, you may also find spiders or other creepy, crawly things in the tiny openings in the chassis or in the power supply. A soft brush, such as a small paintbrush, is good for a quick cleaning, but you can be more thorough with a small vacuum cleaner.

A popular commercial some years ago showed graphically what can happen when a small child decides to "feed" his oatmeal to a VCR. Hopefully, you will never have to clean up this kind of mess.

3.2 Cleaning the Video and Audio Heads

The video drum assembly in a VCR contains the video heads. The heads are located in the uppermost part of this assembly. Most everyone is familiar with the problems caused by dirty video heads. Noise lines begin to appear in the picture and eventually the picture becomes unviewable. Cleaning tapes help VCR owners deal with this problem themselves. When a VCR comes in for servicing, you can do a much better job of cleaning the

Figure 3-3. Cleaning the video heads with a chamois stick pressed up against the drum.

Figure 3-4. Grooves around the upper part of the drum serve as air pockets for the video tape.

Figure 3-5. The diagonal groove on the lower part of the drum assembly can be cleaned with a chamois stick.

heads. What's needed are a chamois stick and head cleaning fluid (usually alcohol based). These are readily available at electronics parts stores and through catalogs.

To clean the head, you spray or dip the chamois stick in the cleaning fluid and press it up against the upper drum assembly as shown in **Figure 3.3**. Then, you turn the video drum around several times by hand. Don't ever move the swab up and down. Just keep it straight on the head in a vertical position with one hand and turn the head with the other hand. Make sure your hands are clean when you perform this operation and don't place your fingers on the surface of the drum. After cleaning, allow a short period of time for the assembly to dry. When you play a tape, you should see a marked difference in the quality of the picture. If not, the video heads may be worn out and need to be replaced.

At the same time you are cleaning the heads, you also are cleaning the grooves around the upper part of the drum (**Figure 3.4**). These grooves provide small pockets of air around the video drum so the tape doesn't adhere to the highly polished aluminum. In effect, the tape rides on a cushion of air assuring that the tape doesn't get stuck to the drum. If there were no grooves, the VCR would not work. The tape would stick as if glued to the highly polished surface. As these grooves become dirty or wear out, increased pressure is put on the tape. It becomes difficult, if not impossible, to adjust the tape tension. These grooves have to be cleaned thoroughly and very carefully. The best way to clean them is with the chamois swabs as you are cleaning the video heads.

One other part of the drum assembly needs to be cleaned. It is the diagonal groove that the tape rests on when in the PLAY or RECORD position (**Figure 3.5**). This groove can be cleaned in the same manner as the grooves in the upper part of the assembly. Place the chamois swab on the groove and turn the drum.

Usually, there are two other head assemblies in a VCR that need periodic cleaning. One is audio/control head assembly and the other is the erase head. These can be cleaned with a cotton or foam swab dipped in alcohol or a head cleaning fluid. These heads are stationary, so you have to move the swab up and down and from side to side to clean off oxide deposits and other contaminants.

The erase head can be found in the tape path, on the left hand side of the video heads as shown in **Figure 3.6**. Naturally, when you are recording a tape, the tape has

Figure 3-6. The VCR's full erase head.

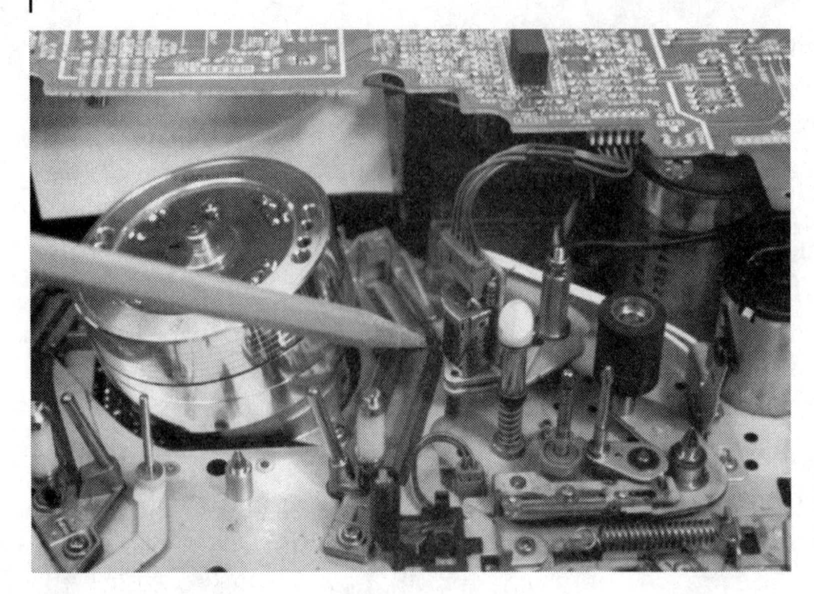

Figure 3-7. **The audio and control heads of the VCR are together on one assembly.**

to be erased before it can be recorded on. This means that the tape must pass over the erase head before it passes over the video heads. The audio and control heads are usually together in one assembly. These heads are located on the right hand side of the video heads as shown in **Figure 3.7**.

3.3 Checking Belts and Tires

All VCRs employ belts and pulleys, some more than others. After many years of operation, belts can lose their spring action. In other words, they stay stretched out when you pull on them (**Figure 3.8**). If you pull on a belt and the elasticity seems to be missing, it is time to change the belt. If you remove a belt from a pulley and set it down on a table, it should form a circle or ellipse. If it forms an odd pattern, such as a peanut shape, you can be sure the belt needs to be replaced.

Belts also can pick up grease and dirt after years of operation. This will cause the belt to slip. A simple remedy is to clean the belt with a cotton or foam swab dipped in a degreasing fluid.

Figure 3-8. **You can check the elasticity of a belt by pulling on it.**

Sometimes, if you examine a belt closely as you stretch it, you will see tiny cracks. This is a classic sign that the belt needs to be replaced. There is no remedy for this.

Usually, when you replace belts, you have to do some disassembly. On some VCRs, particularly those from GE and RCA, you have to remove a whole sub assembly in order to replace two loading belts. Other VCRs are very simple. You don't have to take anything apart, you just replace the belt. It depends on the machine.

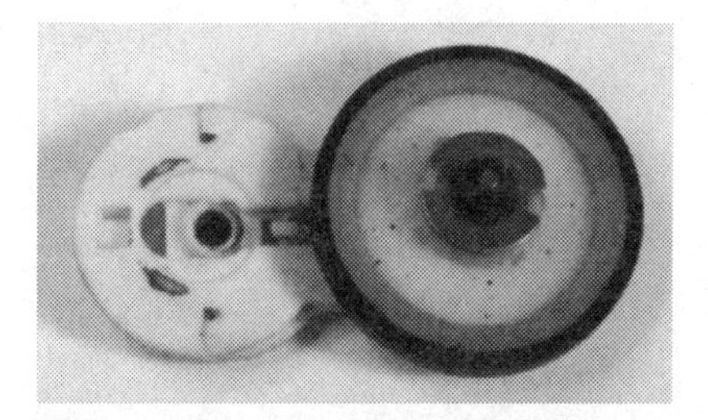

Figure 3-9. The rubber tire on some idler assemblies causes problems when it wears out.

It might seem funny to mention tires when you are talking about VCRs. After all, tires are usually associated with transportation. As far as we know, no VCR uses tires in the same way an automobile does. Instead, tires in a VCR are used to turn wheels through friction. The most popular VCR tire fits around the wheel used in the idler arm assembly (**Figure 3.9**). The job of the idler arm is to turn the reel disks. The idler arm swings to the right reel disk to fast forward a tape and swings over to the left reel disk to rewind a tape. The tire on the idler arm wheel needs to be in good condition to provide the friction needed to turn the reel disks.

Idler arm tires are made of rubber and suffer some of the same kinds of problems as rubber belts. The tires can pick up grease and dirt over time and lose the ability to provide friction for the reel disks. This problem can be solved by cleaning the tires with a foam or cotton swab dipped in a degreasing fluid. Rubber tires also can crack. If you examine the tires closely under a magnifying lamp or glass, you may see lots of tiny cracks. This is an indication that the tire needs to be replaced.

Idler arm tires may also wear down and become too smooth to provide friction. It's not a good idea to clean or sand a worn out tire to increase friction. It's better practice to replace the tire. Sometimes, replacing rubber tires is easy to do. The idler arm assembly comes right out. Sometimes, replacing rubber tires is difficult. Some idler assemblies are positioned in such a way that a great deal of disassembly needs to be done. You'll get a good idea how this is done later in the book.

Chapter 4

Determining The Problem

Good troubleshooting procedure begins with an examination of the external parts of the VCR and then continues with a disassembly of the VCR until you find the cause of the problem. Your powers of observation will greatly help you to determine where the problem lies.

4.1 Checking for Mechanical Problems

If a VCR comes on when you plug it in, you are ready to check for mechanical problems. The first thing to do, before removing the top or bottom covers of the VCR, is to place a tape in the machine and push it in. If the VCR takes the tape, the next thing to do is try to play it. If it plays, then you should try all the rest of the functions. If everything works, the VCR most likely does not have a mechanical problem.

Figure 4-1. Through observation, you can find problems such as this pulley missing its belt.

If you are thwarted in your attempts to load and operate the VCR, you need to remove the top cover and investigate the problem further. With the top cover off, it is easier to see if the problem is mechanical in nature. If the tape will not load into the machine, you may have a problem with the cassette loading assembly. This particular part of the machine has a variety of names depending on the manufacturer. Some names in common use are cage, cassette housing, loading cage, and cage assembly.

A failure to load a tape can be mechanical or electrical in nature. If it is mechanical, a gear may be out of place or a belt may be broken. If the owner tried to force the tape into the machine, metal parts of the loading assembly may be bent or plastic parts may be broken. A good point to start your troubleshooting is to observe the condition of the loading motor. Sometimes, this is a separate unit; other times, the capstan motor does double duty, loading the tape through a series of gears or belts. Try to determine if the motor is spinning and whether or not a belt or gear is engaged. A broken loading belt may be the source of the problem. Again, locating the problem depends on your powers of observation. **Figure 4.1** shows a pulley without a belt. The broken belt caused the loading problem, but the pulley served as a clue to the problem.

If the loading motor seems to be working okay, you should carefully examine all the mechanical linkages of the cassette housing. This is best done with the housing removed from the VCR. To re-

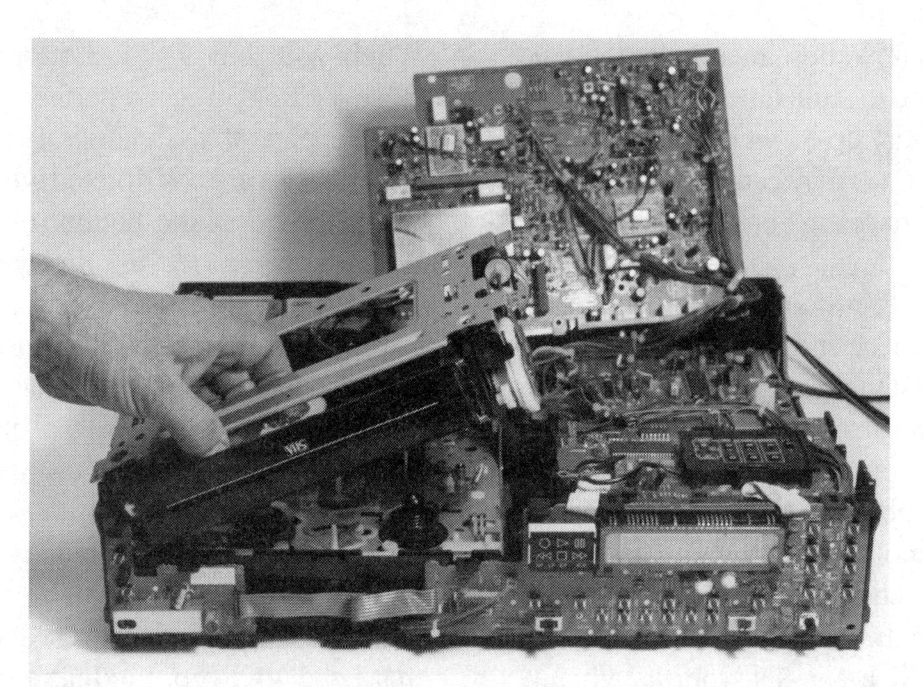

Figure 4-2. For many repairs, you need to remove the cassette housing from the VCR.

move a cassette housing, remove the screws that secure it to the chassis. Then remove any connectors to the main circuit board. **Figure 4.2** shows a cassette housing being removed from a VCR.

Sometimes the housing will not come out easily. You may have to wiggle it around a bit to dislodge it. Pay close attention to the positions of various pieces. Remember, you will have to put the housing back in place eventually. If you can't get the housing out, try loosening the entire chassis. You may have to lift the chassis up slightly to free the housing.

Once you remove the cassette loading assembly from the VCR, you can examine its operation more closely. You do not need a motor to load a cassette tape. You can do it by hand. If you push the tape into the tape holder and manipulate the assembly, you should be able to get the tape to seat properly in the holder. If you can do this, the loading assembly is probably working okay. If not, you need to determine the problem. Again, you should look for gears that are out of line, and bent or broken pieces. If the assembly seems to be beyond repair, you can purchase a replacement. Otherwise, you may want to replace individual parts. If the loading assembly does its job, that is, it loads the tape into the machine and seats it properly, then it is time to check for other mechanical problems.

When you press PLAY or RECORD almost every mechanical part in the VCR springs into action. Gears turn, levers move both on the top and bottom of the VCR. From the top, you can watch as the tape guides pull the tape out of the cassette, slide up the tracks in the chassis, and wrap the tape around the video head drum. When you observe this operation, make sure the tape is moving and that the tape guides are sliding all the way up to their final position. Any broken links may cause the tape guide to stop prior to seating at the end of its run.

Another part to watch closely is the pinch roller. This piece should swing into place firmly against the capstan shaft. The pinch roller and capstan shaft work together to pull the tape out of the cassette. The capstan shaft is powered by the capstan motor and, thus, should be turning. The rubber pinch roller should provide enough friction to move the tape. **Figures 4.3** and **Figure 4.4** show a chassis from a Sylvania VC4243AT01 before and after PLAY or RECORD is pressed. If you compare the two figures closely, you will see how the tape guides, pinch roller, and other parts move in the chassis.

A test jig is a very useful tool for figuring out if the mechanical parts of the VCR are working properly. It is the size of a cassette tape, but is made of clear plastic When you place it in the VCR, the machine thinks a real tape has been loaded. You get a birds-eye view of all the mechanisms usually hidden underneath the cassette tape, such as the idler arm assembly. Plus, you can see all the mechanisms move when you press a key such as PLAY.

For a thorough check of the mechanical linkages, you have to look underneath the chassis. You may think that this is simply a matter of removing the bottom cover of the VCR. But this is not always the case. Some VCRs are constructed with the main circuit board at the top of the machine. When you remove the bottom cover of these VCRs, you get a perfect view of the bottom part of the chassis (**Figure 4.5**). Other VCRs are constructed with the main circuit board at the bottom of the VCR. When you remove the bottom cover of these VCRs, all you can see are small portions of the circuit board (**Figure 4.6**). To examine the bottom of the chassis in these VCRs, you have to remove the chassis from the VCR case and turn it upside down.

No matter how the VCR is constructed, the bottom of the chassis contains most of the pulleys, gears, springs, levers and belts that make the VCR go. Finding a mechanical problem often demands patience and the ability to figure out how all the mechanisms work. Another clue to mechanical problems are the noises the VCR makes when it is playing. You may hear the whirring noise of a motor or the grinding sound of a gear. Any strange noise is an indication that one of the VCR's mechanisms is not functioning properly. For example, sometimes you'll hear noise coming from the spinning video heads. Usually, this noise is caused by bad bearings in the drum motor.

4.2 Checking for Electrical Problems

When you plug in a VCR, you should immediately check the display to see if it is blinking. If there is no display, then you know you have an electrical problem. Most of the time, this indicates a problem in the power supply. Some older VCRs employ a DC-to-DC converter (RCA and Hitachi), which produces a DC voltage for the dis-

Figure 4-3. A Sylvania VCR chassis before pressing PLAY.

Figure 4-4. The same chassis after pressing PLAY.

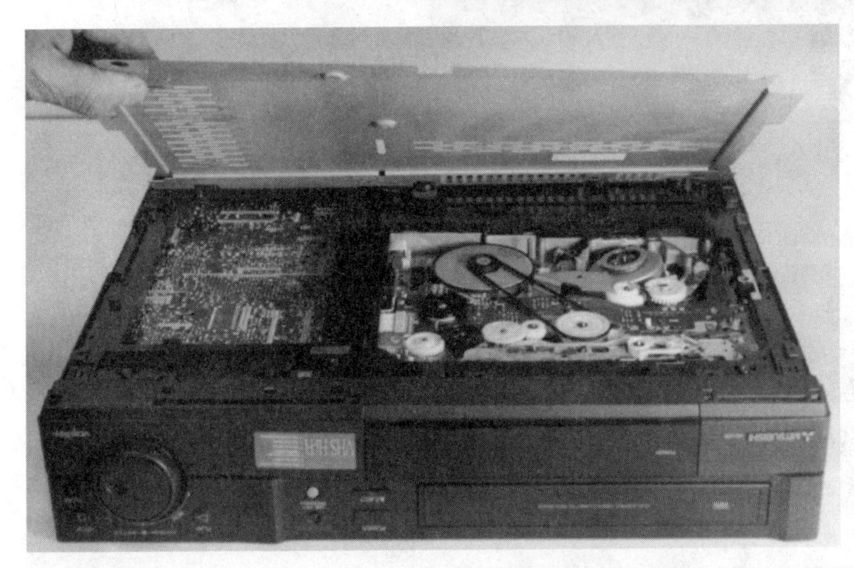

Figure 4-5. Removing the bottom cover of some VCRs gives you a perfect view of the mechanical components.

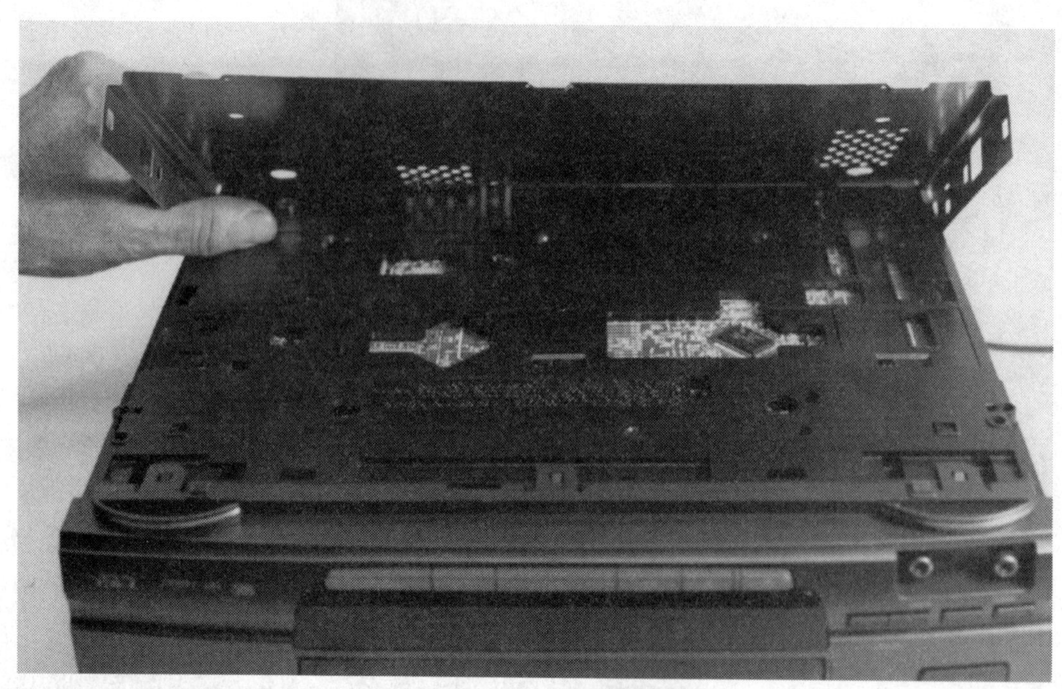

Figure 4-6. On some VCRs, removing the bottom cover is useless.

Figure 4-7. The bottom of some reel disks have a pie shaped pattern, used for the VCR's counter.

play and the tuner. If the display doesn't work, then the tuner section of the VCR doesn't work.

Another indication of an electrical problem is erratic operation of the loading mechanism or the VCR functions. Sometimes, when you load a tape, it starts to go in, then it comes out again. This often indicates a defective sensor or broken connection leading to the sensor. Sometimes, when you press PLAY, the VCR ejects the tape or goes into reverse. This "mixed up" behavior often points to a defective or dirty mode switch.

Sometimes, the VCR will play, but the counter will not count. Then, after several seconds, the VCR shuts off. This can be caused by a defective sensor under the right or left reel disk. On some VCRs, if you lift up the reel disk and turn it over, you will see a pie-shaped pattern (**Figure 4.7**). This pattern works in conjunction with a special sensor to send pulses to the counter circuits in the microprocessor.

If the customer complains about poor video, the problem may be caused by the video heads, tracking circuits, or power supply. Connecting the VCR to a monitor helps you determine the source of the problem. If there are lines all over the picture, the video heads may need to be cleaned or replaced altogether. If there are lines in one area of the picture, there may be a problem with the tracking circuits. If there is a band of interference in the middle of the picture, it is probably due to a faulty capacitor in the power supply producing 60-cycle hum.

Sometimes, you may notice that the video head does not spin at a constant rate, but instead slows down, increases, and slows down again. This is an indication of problems in the servo circuits, since these circuits control the speed of the motor.

Electrical problems often occur when the printed circuit board itself has problems. Typical PCB problems are cracked solder joints, broken traces, or burned wires or components. These problems can be found by close inspection of the PC board with a magnifying lamp or glass.

4.3 Checking for Problems Caused by Abuse

If a VCR has been dropped or stepped on, it usually suffers physical damage outside and inside the machine. On the outside the cover may be bent or the front panel broken. On the inside you may find a cracked printed circuit board or broken components. Usually, all that is needed to repair these physical problems is to patch the cracks in the board and replace the components.

Chapter 5

Troubleshooting Mechanical Problems

VCRs are complex mechanical machines employing gears, pulleys, springs, levers, belts, brackets, wheels and other parts, all driven by small motors. Repairing the mechanical parts of a VCR certainly demands mechanical aptitude. You have to know how to take assemblies apart and how to put them back together. In this chapter, we try to help you understand what can go wrong mechanically in a VCR.

5.1 Checking Gears, Belts, Tires, Springs and Bumpers

A visual inspection of the VCR will show if gears are worn out, cracked, or have broken teeth. A broken gear is obvious (**Figure 5.1**). Pay special attention to gears that are mounted on a shaft. Sometimes they crack near the shaft and become loose. Then, the gear turns but the shaft doesn't. This is especially true with JVC VCRs. These machines have a small toothed gear mounted on the capstan motor. This gear has a very thin wall and develops cracks over time. Then, even though the capstan motor turns, belts and pulleys that rely on the gear do not turn. This is because the gear does not have a grip on the capstan motor shaft. You cannot glue this, or repair it, you have to replace the gear with a new piece.

There are two types of belts in a VCR. One is a square belt made from rubber, the other is a toothed belt made from nylon (JVC, Panasonic, Mitsubishi, and certain other models use toothed belts). To check a square rubber belt you have to stretch it a little to see if the belt is cracked or has lost its elasticity. If it is cracked or loose it must be replaced. Another way to check the belt

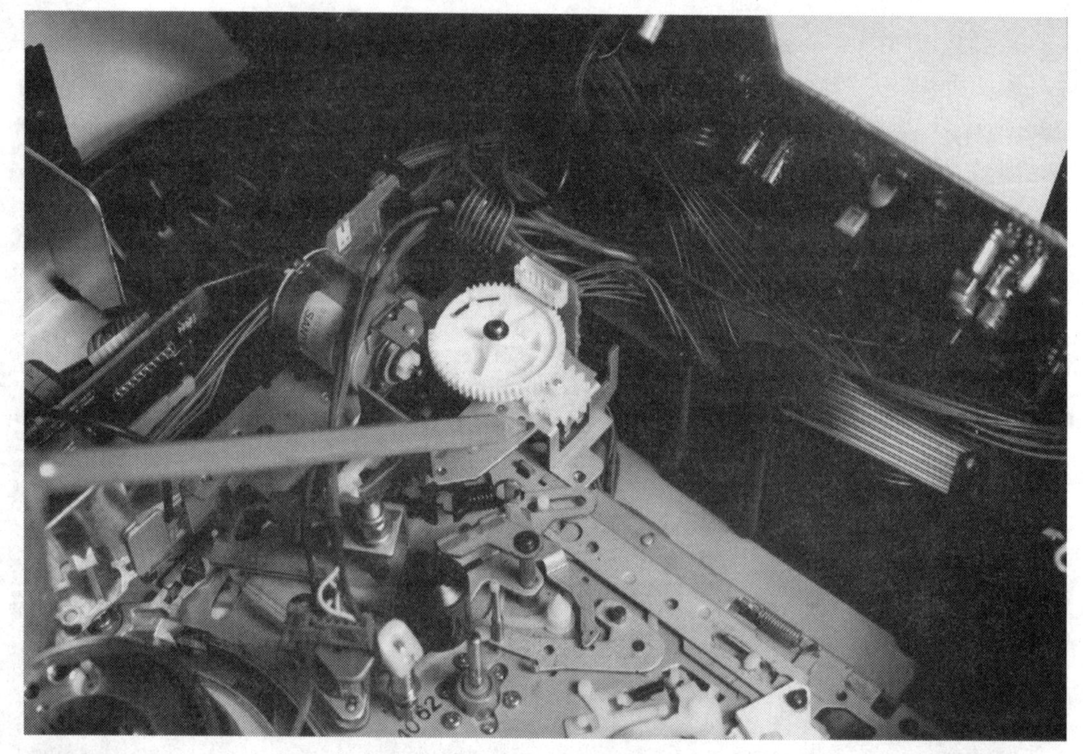

Figure 5-1. Through visual inspection, you can locate problems such as this gear, which has a broken tooth.

is to hold the wheel that it is turning. If you feel a good resistance, the belt is good. If there is very little resistance, the belt needs to be changed.

Toothed belts are different. They are not elastic, but tend to enlarge slightly over time. Most VCRs that employ toothed belts have a provision for adjusting the tension on the belt. Typically, this is a small wheel with a bracket mounted by a single screw. When you loosen the screw you can move the wheel and thus increase the tension.

When replacing belts, there is a device you can use to check the length of the old belt against the length of the replacement belt. You place the old belt over two of the shafts to take a measurement. The new belt should measure slightly smaller than the old belt (**Figure 5.2**).

Tires are mounted on wheels. On some early VCRs, the reel disks had rubber tires, and the idler arm assembly wheel did, too. In later models, only the idler arm assembly wheel had a tire. In recent years, no rubber tires are used, toothed gears are used instead. Depending on the VCR you are servicing, you have to check the tires very carefully because they are a source of many problems. Over the years, tires either wear out or harden, even if the VCR is not in use. The symptom of a bad tire is slipping. A visual inspection will show that the tire is shiny. The tire can be removed from the wheel and replaced with a new one. Trying to fix the tire by sandpapering usually doesn't work at all or is only a temporary fix.

One of the ways to remove a tire from a wheel is to slip a very small flathead screwdriver between the tire and the wheel. Then, you have to pry the tire off the wheel. In some cases, you have to remove the wheel from the VCR before performing this procedure. Other times, you can leave the wheel in place. Usually, you can place a new tire on a wheel by hand. You should check your work carefully, though, to make sure the tire is seated properly.

Springs cause problems when they detach or suffer metal fatigue. If a spring is detached, you should check why this happened. Maybe the end of the spring is broken or the spring has loosened and doesn't hold in the holes anymore. If the spring is broken it needs to be replaced with a new one.

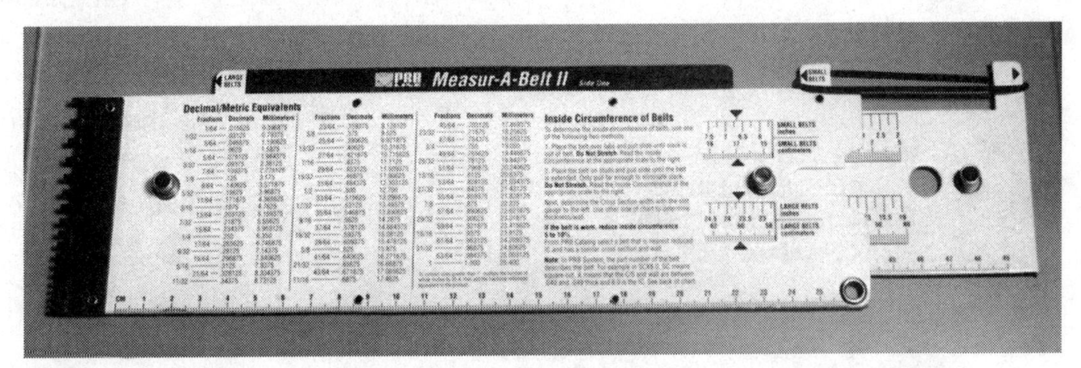

Figure 5-2. A belt measuring tool can help you determine the correct replacement belt.

On some models, especially Fisher VCRs, there is a spring attached to the idler arm assembly. The tension in this spring is very important to the functioning of the idler assembly. Because the spring is located at the top of the chassis, sometimes it jumps out of the hooks that hold it and falls inside of the VCR. The machine will not function if this happens. It will shut off. If you notice the spring is missing, turn the VCR upside down and shake it. This is the good way to find the spring. Then you can reinstall the spring, making sure the hooks are not damaged.

Springs engage levers, wheels, and gears and thus are very important to the operation of the VCR. Make sure to pay close attention to the springs.

Rubber bumpers are used in some Funai VCRs (also Sylvania and Teknika). These unassuming pieces of rubber, about one-quarter inch high, play a big role in the fast forward and rewind of these VCRs. They are located on metal tabs to the right of the idler arm assembly. Over time, the bumpers become compressed or break apart **(Figure 5.3)**. Replacing a bumper is just a matter of removing the old one and slipping on a new one.

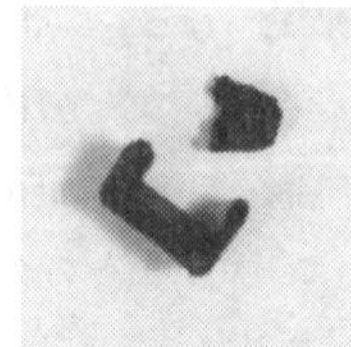

Figure 5-3. A broken rubber bumper can affect the fast forward and rewind operations of some VCRs.

5.2 Checking the Cassette Loading Mechanism

There are two types of cassette loading mechanisms. Very early VCR models used top loading manual mechanisms. These are now obsolete. The loading mechanism in use today is the front loading type, which is powered by an electrical motor. Some VCRs employ a separate motor for the loading mechanism, controlled by the microprocessor. Other VCRs make the capstan motor do double duty, powering the loading mechanism through an arrangement of gears, wheels, and belts. This design saves one motor.

The cassette loading mechanism typically contains three leaf switches and two tape end sensors. When you push the tape into the VCR, the switch mounted on the top of the loading housing turns the power to the loading motor on. As the motor turns and the tape is loaded, a worm gear engages a switch that stops the loading motor from turning. When you press EJECT, the loading motor turns in reverse and the same worm gear engages another switch that stops the motor when it comes to the end of the cycle.

All these gears are made from plastic and tend to flex if there is any problem in the cassette housing. This flexing can bend the leaf switches and cause the motor to continue running after it comes to the end of the cycle. This will shut off the power supply.

If you encounter a problem with the loading mechanism, you have to pay particular attention to the leaf switches **(Figure 5.4)**. They should not be bent or broken. In case there is a problem, you can remove the cassette housing, disconnect the plug attached to the power supply, and do the necessary inspection and repairs.

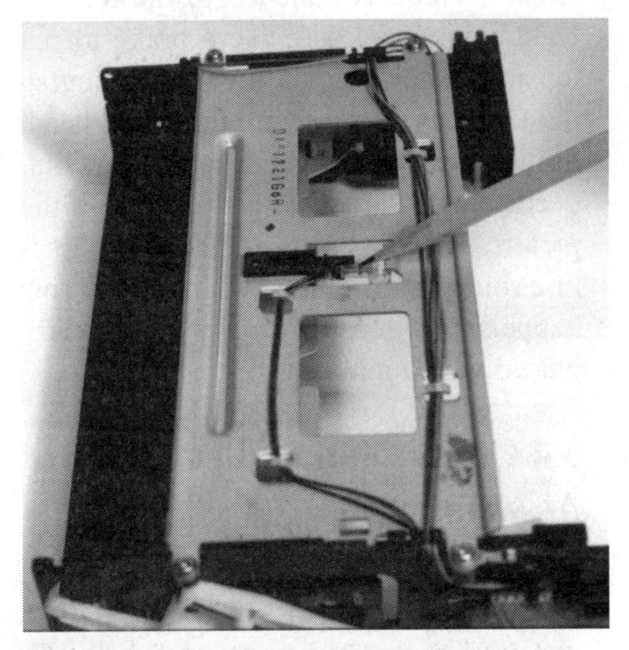

Figure 5-4. Some cassette housings employ leaf switches, which you should examine carefully for proper operation.

A cassette loading mechanism without its own motor is a completely different story. This type is controlled by a mode switch. There are no switches or sensors on the loading mechanism. It is purely mechanical. Problems with this type are broken plastic gears or broken toothed rails, depending on the model. Also, problems arise when the belt that goes to the capstan motor wears out. You must pay close attention to the loading belt.

Some VCR loading mechanisms use a loading motor mounted on the chassis. The mechanism takes power from the motor through a series of belts and gears.

5.3 Checking the Tape Path

You can check the tape path very easily by loading a tape and pressing the PLAY button (**Figure 5.5**). You need to pay close attention to certain points. First is the pinch roller assembly, the most troublesome spot in the tape path. You have to check that the tape goes in straight and comes out straight.

Figure 5-5. To examine the tape path, press PLAY and watch closely.

The second point is the audio head. The tape must pass over the audio head without touching the shield of the head. The tape has to be smooth, not bent or curled. This way, the audio head reads the signal on the tape properly. On some VCRs, there is a small tape guide that is mounted between the pinch roller and the cassette. This guide sometimes bends due to users jamming the tape into the machine. A bent guide will damage the tape. The guide changes the tape's height and perpendicularity to the chassis. This guide can be straightened but the best thing to do is replace it with a new one.

The third point is the outgoing tape guide next to the video head. Over time, the tape guide accumulates oxides and makes the guide smaller from top to bottom than the width of the tape. This causes erratic speed control. This guide has to be cleaned very carefully and very gently with a chamois swab, the same as the one used to clean the video head.

The fourth point is the incoming tape guide, which is the same type as the outgoing. The fifth point is the erase head, which has its own guide. On some machines, the erase head is mounted on the chassis, on others there is a so-called flying erase head, which is mounted on a spring-loaded shaft. This head is not stationary. To the left of the head there is a tape guide that adjusts the height of the tape relative to the erase head. If there is a problem with the tape curling, you can align this manually.

The tape tension guide is the sixth and final point of the tape path and the most important point. The tape tension guide is a round pin mounted on a bracket. The tension is provided by a small spring. This tension guide is attached to the tension band. This band acts like a brake to put tension on the tape through the spring so the tape has contact with the video head. Sometimes the felt on the band wears out, decreasing the tension and distorting the picture. This distortion may result in a picture rolling and covered with noise. If this happens, the tension band should be replaced with a new one.

5.4 Checking the Idler Assembly

The idler arm assembly **(Figure 5.6)** transfers power from a vertically grooved plastic wheel (turned by the capstan motor) to the reel disk. The idler assembly is a key component in the fast forward and rewind

Figure 5-6. The idler assembly is a key component in the fast forward and rewind operations of some VCRs.

operations. As mentioned, in older models idler assemblies use rubber tires. These tires can become hardened or worn and lose their grip. Replacing the tire with a new one will restore its function.

On newer VCRs, the idler arm assembly uses a toothed gear. Sometimes these gears are broken by excessive force (forcing a tape into the VCR). If you see broken teeth, the gear has to be replaced.

Most of the time, removing an idler arm assembly entails removing the cassette housing and a few other parts. In some VCRs, though, this assembly can be removed quite easily, just by removing one screw.

If a VCR will not fast forward or rewind, but operates normally otherwise, the problem may be caused by a worn rubber tire on the idler assembly. But this depends on the VCR model. Some VCRs with the same symptoms have a worn out rubber bumper.

5.5 Checking the Mode Switches

Two types of mode switches are used in VCRs. Early models use a sliding mode switch; newer models use a multi-position rotary switch. Either way, the function is the same.

The mode switch **(Figure 5.7)** signals the microprocessor regarding the status of the mechanical assembly at every moment. If you want to eject a tape while it is fast forwarding, the mode switch signals the microprocessor that the tape is fast forwarding. The microprocessor stops the tape first and then ejects it.

For the mode switch to work properly, it must be correctly aligned with the whole mechanical assembly. The position of the mode switch is usually signified by a dot on the wheel and a mark on the body of the switch. The proper alignment of each VCR depends on the manufacturer. Some-

Figure 5-7. The mode switch signals the microprocessor about the status of the mechanical assembly.

times the dot and mark come together when the cassette is out of the machine. But this is not the case in all machines. If you are not sure how the mode switch should be aligned, you have to refer to the service manual.

Many times the problem with the mode switch is dirty contacts. This can create all sorts of problems. The symptom is an erratically functioning VCR. You press fast forward and the tape doesn't move, or the power shuts off. Or, while the machine is playing, it starts fast forwarding by itself.

You can clean the mode switch by spraying it with WD-40 lubricant. Spray the switch, press PLAY to turn it, spray it again, and press PLAY again. Spray the mode switch from the side. If you remove the mode switch from the VCR, you can turn it manually. Sometimes the mode switch is not accessible and you have to disassemble the VCR to get to it, as in certain Sharp VCRs.

5.6 Mechanical Adjustments

Mechanical adjustments can be done by hand, but it takes some experience. If you are not successful, you have to purchase specialized tools such as a tension meter. However, these are expensive, costing several hundred dollars each.

We'll explain how to do adjustments manually. But, be advised that mechanical adjustments are not needed in most repairs. Suppose a problem, such as tape curling, occurs due to a defective pinch roller. All you need to do is replace the pinch roller. Normally, you don't need to do any adjustments. In most cases, even if you replace the video head, you don't need to do any mechanical adjustments.

Mechanical adjustments are needed when the tension band **(Figure 5.8)** is broken and you replace it with a new one. Then, you have to do a tape tension adjustment. This involves adjusting the tension on the

Figure 5-8. A broken tension band.

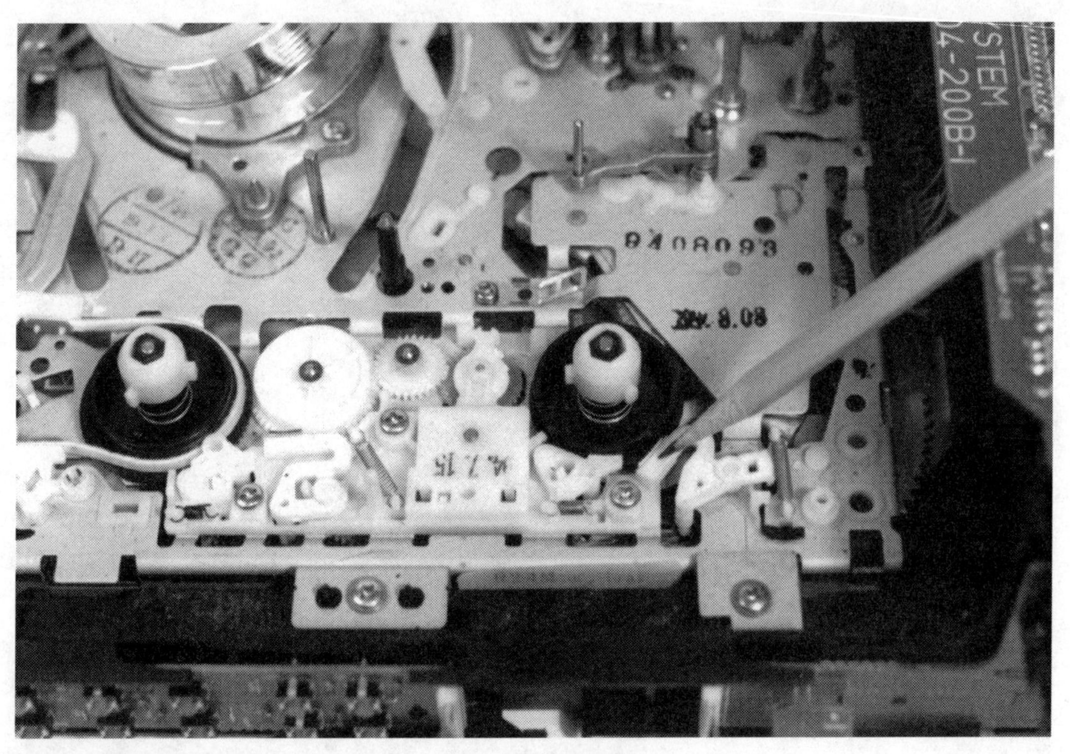

Figure 5-9. The brakes can be adjusted if the VCR allows it.

spring. On one side, the spring is attached to a small bracket with a rectangular hole inside. By moving it up and down, you adjust the tension on the spring and thus on the tape. This adjustment can be done without a meter, but you have to have a feel for it. It cannot be too loose or you will lose the top half of the picture. If it is too tight, the video head will wear out prematurely and may produce a jittery sound. One of the ways to make the adjustment is to loosen the spring until you lose the top half of the picture or the picture becomes shaky. Then, incrementally increase the tension on the spring, while observing the picture quality. Stop at the point where you see a good picture.

The brakes **(Figure 5.9)** sometimes need adjustment, if the machine allows it. If not, you must change the brake pads. The basic rule for adjusting the brakes is this: With the VCR in STOP mode and with the tape out of the machine, turn the right reel disk to the right until the resistance is equal to turning the left reel disk to the left.

Chapter 6

Troubleshooting Tuner/ Demodulator Circuits

The tuner/demodulator in a VCR is typically housed on a separate circuit board and placed in a shielded enclosure. The entire unit is soldered in perpendicular to the main circuit board. **Figure 6.1** shows the tuner/ demodulator section of a GE VG-7725 VCR. The schematic diagram of this section is shown in **Figure 6.2.** In this VCR, the tuner is a varactor tuner, found in many modern VCRs.

If you desolder the tuner/demodulator from the VCR and remove the shields on either side, you can see the capacitors, coils and ICs that comprise this unit. **Figure 6.3a** shows one side of the device; **Figure 6.3b** shows the other. You will notice that the foil side of the tuner/demodulator is very densely packed with surface mount components.

If the symptoms of the VCR point to the tuner, such as snow in the picture or weak reception, the best course of action is to replace the entire tuner, which costs about $60. In fact, service manuals routinely note that FCC specifications will not be satisfied if components of a UHF/VHF tuner and frequency synthesizer sections are replaced individually. But, replacing the tuner doesn't always clear up the picture. This can cause anger and frustration, since this is such an expensive part to replace. Before you decide to replace the tuner, you should examine it carefully under a magnifying lamp for cracked solder joints at the points where it connects to the main circuit board. Resoldering the joints can sometimes solve the problem.

Figure 6-1. The tuner/demodulator section is protected by a metal shield.

Figure 6-3a. The component side of the tuner/demodulator section with the shield removed.

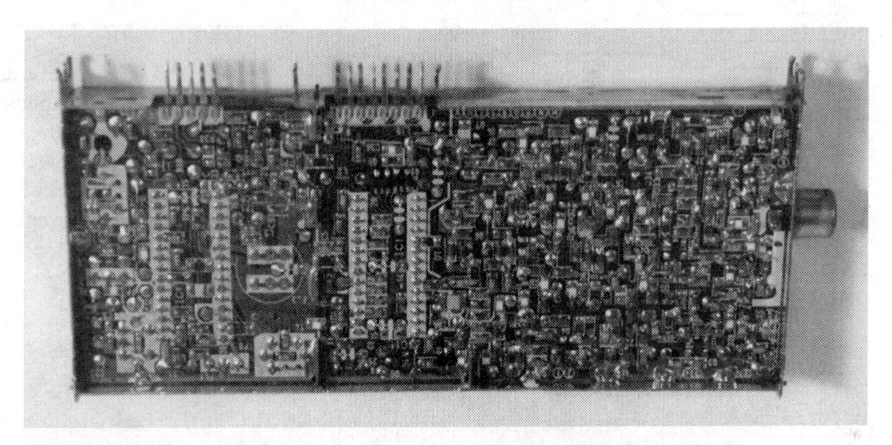

Figure 6-3b. The foil side of the tuner/demodulator section with the shield removed.

6.1 The Tuner/Demodulator

The signal that comes to the tuner from the antenna or cable is very weak. The tuner mixes the frequency of the signal with the local oscillator and produces an intermediate frequency signal, which is also very weak, about 3 mV.

From this point, to achieve a clear snow free picture, the tuner has to amplify this signal. This is done in three stages, first IF, second IF and third IF. In this tuner, the first IF is discrete circuitry, which consists of one transistor (2SC4417) and its supporting circuitry. This amplifies the signal about 20 to 25 dB (10 times). After amplification, the IF signal is fed to a SAW (surface acoustic wave) filter. This a ceramic buffer type filter, which is factory adjusted to the IF frequency (45.75 MHz). The filter also has about 25 dB losses, so the signal returns to its original low level. Thus, the first IF can essentially be regarded as a driver for the SAW filter.

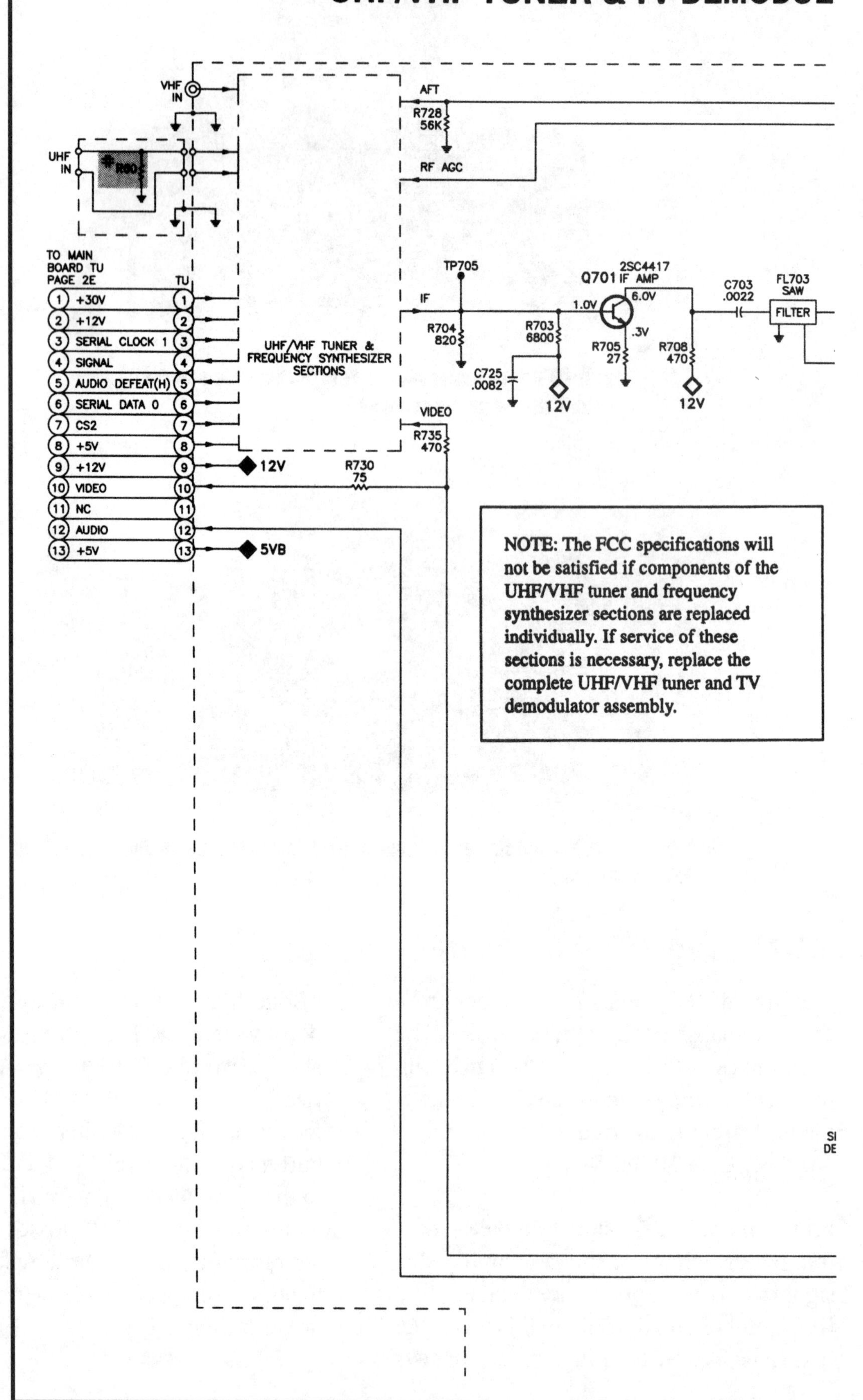

Figure 6-2. The schematic diagram of the tuner/demodulator section.

ATOR ASSEMBLY SCHEMATIC

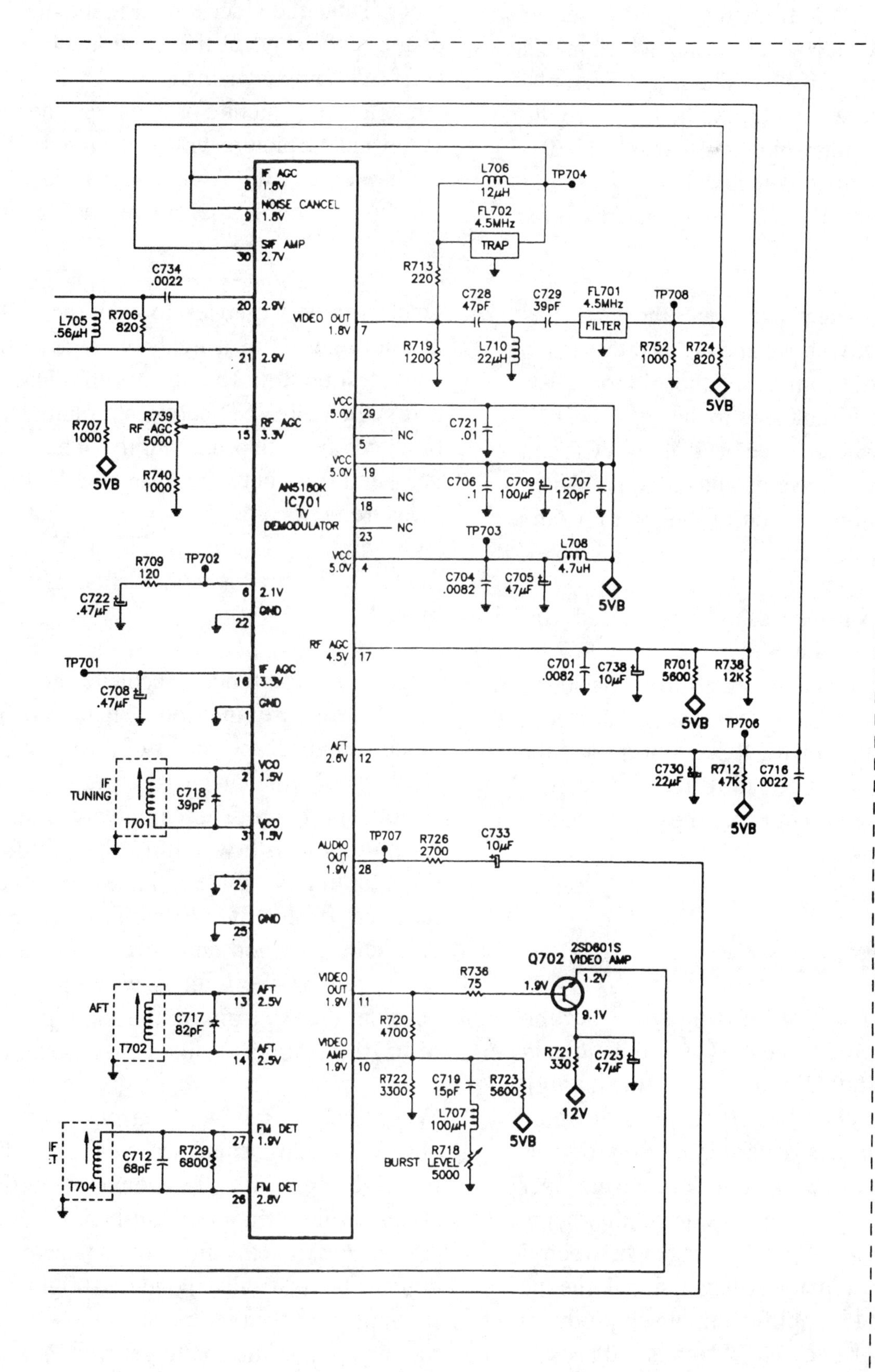

From the SAW filter the signal is fed to the second video IF amplifier, which in this case is part of IC701. The signal is amplified by the second IF and then goes to the video detector, which is also part of IC701. From the video detector the signal is fed to the sound IF filter (FL701). This filter is external to the IC.

The video detector separates the video portion of the IF signal and sends it to a video amplifier. From here, the video signal leaves the IC and goes to the VIDEO OUT RCA connector at the back of the VCR and also to the rf converter and through a switch (an electronic switch in this case), to the recording circuits.

The audio also needs to be processed. Audio goes to the FM detector of IC701, which separates the sound from the sound IF (SIF). The audio leaves IC701 and goes to the AUDIO OUT RCA connector at the back of the VCR and to the rf converter and through a switch, to the audio recording circuits.

6.2 AFT and AGC Circuits

Also part of the tuner/demodulator section of the VCR are the AGC (automatic gain control) and AFT (automatic fine tuning). When the video detector detects the video signal from the IF carrier, it also produces a DC voltage, which is fed back to the tuner and the IF amplifiers to control their gain. This voltage typically ranges between 7 and 10 V. When a stronger signal hits the tuner and IF amplifier circuits, it produces a higher DC voltage, which is fed back to the tuner and changes the sensitivity of the tuner. Thus, the video signal at the output of the tuner stays at the same level. Any defective component in the AGC will cause either a snowy picture or will overload the video and the picture will be saturated. The components of the AGC circuits must be carefully checked before replacing the tuner.

One commonly used test to see if the AGC is working is to measure the voltage at the AGC test point of the tuner with a DMM and switch the VCR between a blank station and one with programming. The voltage should be between 7 and 10 V. Any measurement below 5 V indicates that the AGC circuit is not working. In most cases, the IC (IC701 in this model) should be replaced.

In order for the tuner to achieve the best possible picture and sound it has to be tuned exactly at the center of the station's frequency. With manual, rotary tuners, this is not a problem. You can always adjust the tuning while watching the picture. Varactor tuners, however, have a circuit to take care of this. It is called the AFT circuit. One common and very easy test to tell if the AFT circuit is working is to desolder the AFT pin (in this case, pin 12 of IC701) from the printed circuit board.

When the VCR is working on a certain channel, touch a small screwdriver to the pin of the IC so it makes contact with the traces on the printed circuit board. If the AFT is working, the moment you make the connection, you will experience a slight improvement in the picture and the sound. If the picture and the sound get worse, it indicates that the AFT is not working.

To disconnect one pin of an IC, use either a solder wick or a desoldering tool. Once the solder is removed from the hole, just make sure the pin is not making contact with the hole or the solder around it.

6.3 The Frequency Synthesizer

One part of the tuner we have not touched upon yet is the frequency synthesizer IC (not shown in the diagram but part of the tuner section). In the photo of the disassembled tuner, you can see it as IC7601. **Figure 6.4** shows the pinouts of this IC. The frequency synthesizer controls the tuner, changes the channels, and adjusts the AGC and AFT.

The frequency synthesizer tuning system is controlled by the microprocessor. The microprocessor receives the channel select (channel up/down) signal and channel information from the front panel buttons or from the infrared receiver. The microprocessor then sends the information to pin 23 of the frequency synthesizer IC in serial form. This IC then converts the data to parallel form and presents it as output at pins 1 through 4. This information is used by the tuner to tune in the channel.

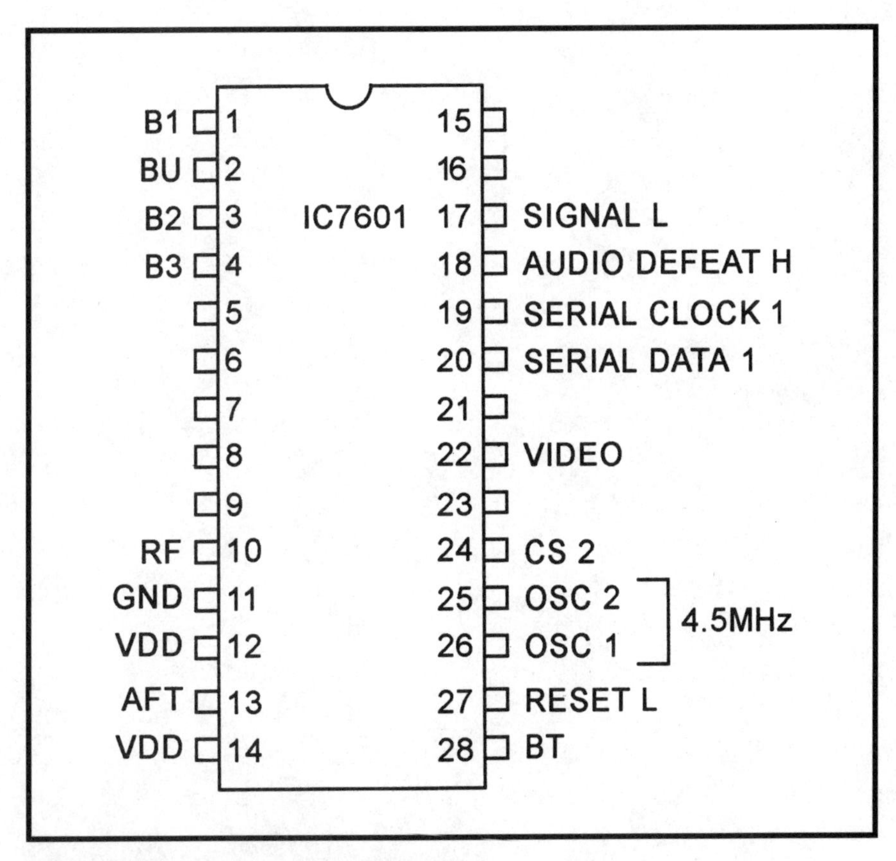

Figure 6-4. The frequency synthesizer IC.

Chapter 7

Troubleshooting Video Circuits

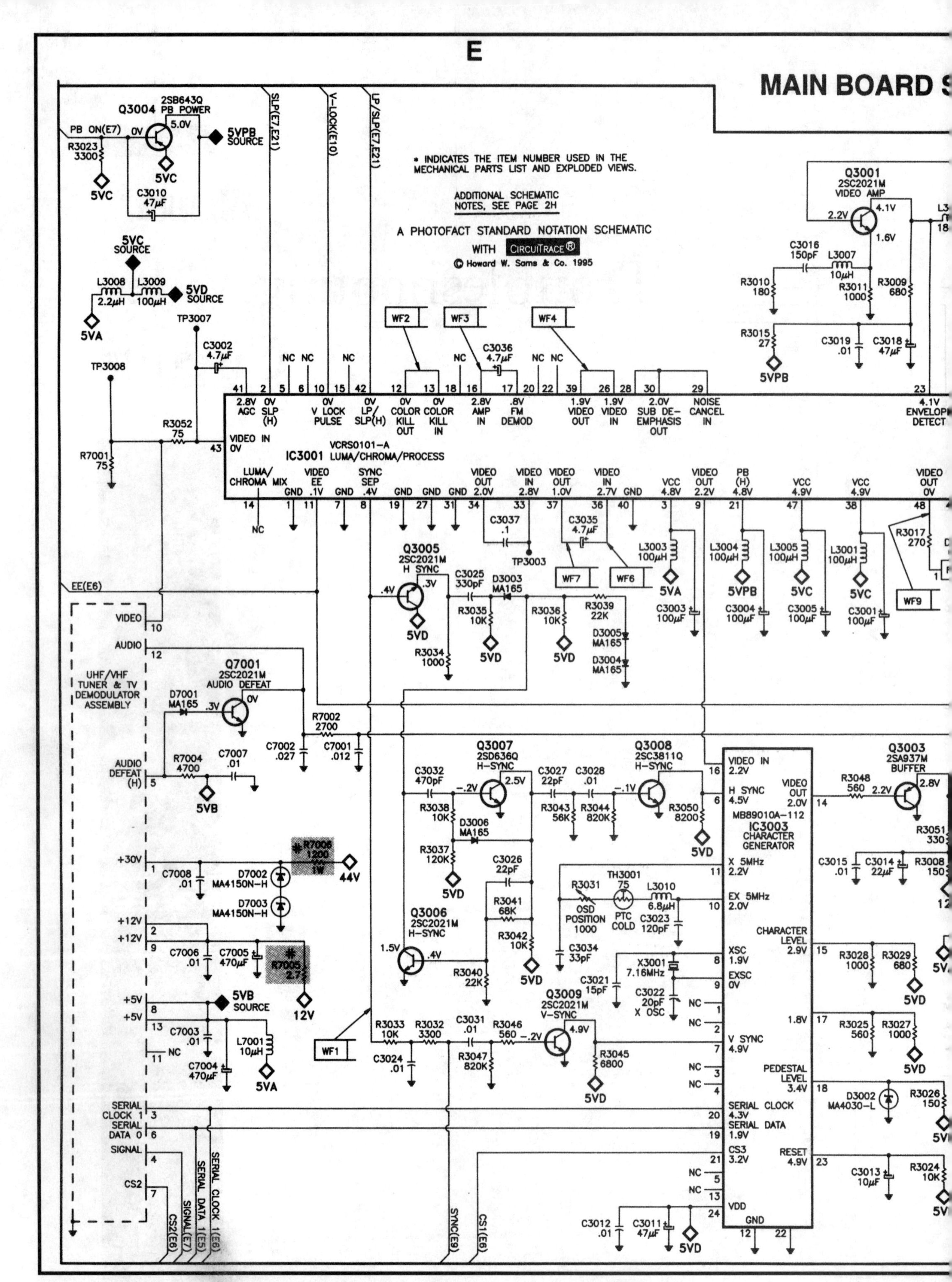

Figure 7-1. Schematic diagram of the video section showing IC3001, the Luma/Chroma Processor.

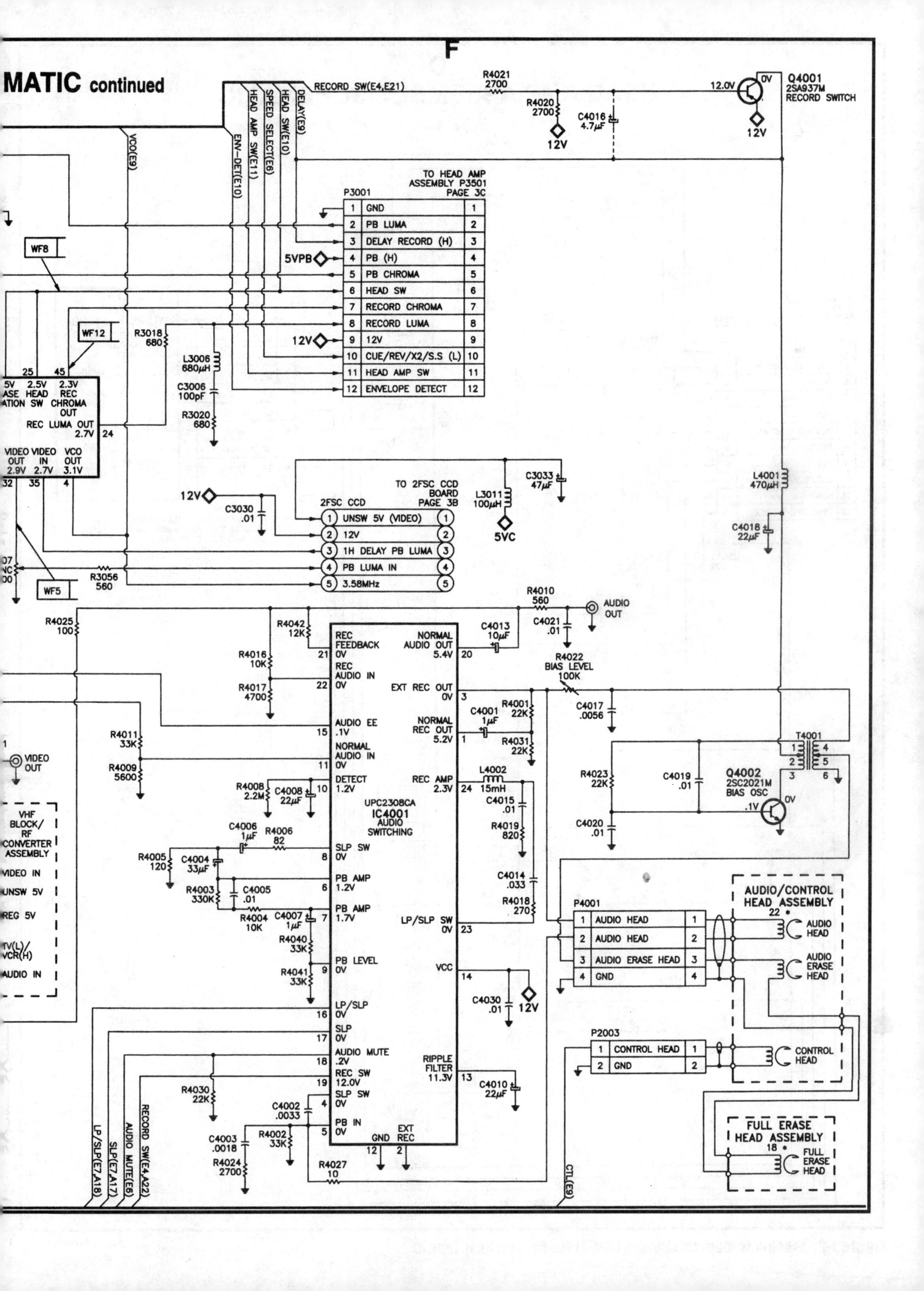

F
MATIC continued
RECORD SW(E4,E21)
R4021 2700
Q4001 2SA937M RECORD SWITCH
12.0V
0V
R4020 2700
C4016 4.7µF
12V
TO HEAD AMP ASSEMBLY P3501 PAGE 3C
P3001
1 GND
2 PB LUMA
3 DELAY RECORD (H)
4 PB (H)
5 PB CHROMA
6 HEAD SW
7 RECORD CHROMA
8 RECORD LUMA
9 12V
10 CUE/REV/X2/S.S (L)
11 HEAD AMP SW
12 ENVELOPE DETECT
5VPB
VCO(E9)
ENV-DET(E10)
HEAD AMP SW(E11)
SPEED SELECT(E6)
HEAD SW(E10)
DELAY(E9)
WF8
WF12
WF5
R3018 680
L3006 680µH
C3006 100pF
R3020 680
REC LUMA OUT 2.7V
VIDEO OUT 2.9V
VIDEO IN 2.7V
VCO OUT 3.1V
TO 2FSC CCD BOARD PAGE 3B
2FSC CCD
1 UNSW 5V (VIDEO)
2 12V
3 1H DELAY PB LUMA
4 PB LUMA IN
5 3.58MHz
C3030 .01
C3033 47µF
L3011 100µH
5VC
R3056 560
L4001 470µH
C4018 22µF
R4010 560
AUDIO OUT
C4021 .01
R4025 100
R4042 12K
C4013 10µF
R4016 10K
R4017 4700
R4022 BIAS LEVEL 100K
C4017 .0056
R4001 22K
C4001 1µF
R4031 22K
R4011 33K
R4009 5600
VIDEO OUT
T4001
Q4002 2SC2021M BIAS OSC
R4023 22K
C4019 .01
C4020 .01
R4008 2.2M
C4008 22µF
L4002 15mH
C4015 .01
R4019 820
UPC2308CA IC4001 AUDIO SWITCHING
REC FEEDBACK 0V
REC AUDIO IN 0V
AUDIO EE .1V
NORMAL AUDIO IN 0V
DETECT 1.2V
SLP SW 0V
PB AMP 1.2V
PB AMP 1.7V
PB LEVEL 0V
LP/SLP 0V
SLP 0V
AUDIO MUTE .2V
REC SW 12.0V
SLP SW 0V
PB IN 0V
NORMAL AUDIO OUT 5.4V
EXT REC OUT 0V
NORMAL REC OUT 5.2V
REC AMP 2.3V
LP/SLP SW 0V
VCC
RIPPLE FILTER 11.3V
GND
EXT REC
VHF BLOCK/ RF CONVERTER ASSEMBLY
VIDEO IN
UNSW 5V
REG 5V
TV(L)/ VCR(H)
AUDIO IN
C4006 1µF
R4006 82
R4005 120
C4004 33µF
R4003 330K
C4005 .01
R4004 10K
C4007 1µF
R4040 33K
R4041 33K
C4014 .033
R4018 270
P4001
1 AUDIO HEAD
2 AUDIO HEAD
3 AUDIO ERASE HEAD
4 GND
AUDIO/CONTROL HEAD ASSEMBLY
AUDIO HEAD
AUDIO ERASE HEAD
C4030 .01
P2003
1 CONTROL HEAD
2 GND
CONTROL HEAD
C4010 22µF
R4030 22K
C4002 .0033
C4003 .0018
R4002 33K
R4024 2700
R4027 10
FULL ERASE HEAD ASSEMBLY
FULL ERASE HEAD
LP/SLP(E7,A18)
SLP(E7,A17)
AUDIO MUTE(E6)
RECORD SW(E4,A22)
CTL(E9)

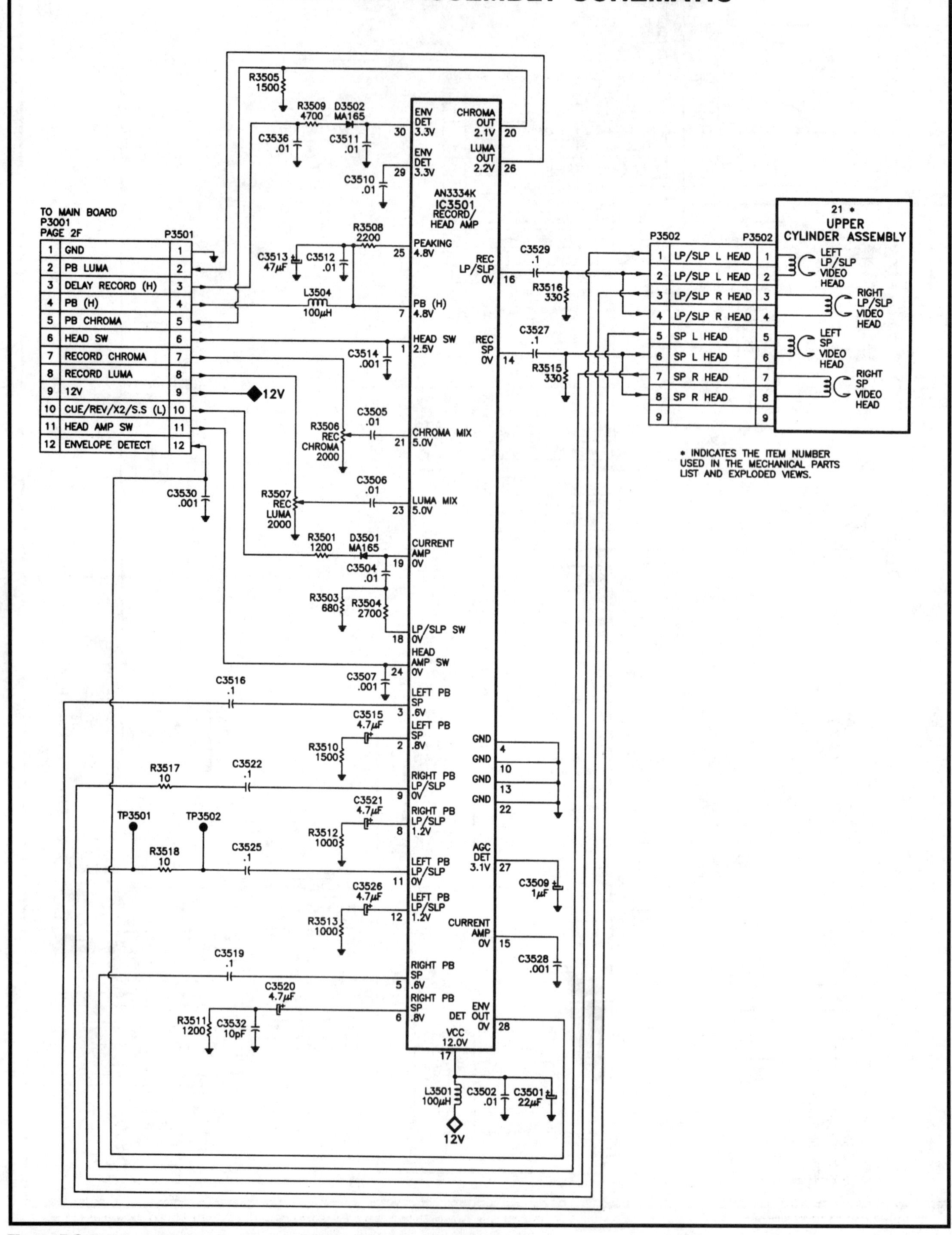

Figure 7-2. Schematic diagram showing IC3501, the Record/Head Amp IC.

Poor video performance in a VCR is most often caused by dirty or worn video heads. If there is any indication that the video is working properly, such as good video output at a slow speed but poor video output at a fast speed, you can be fairly certain the problem lies in the video heads. Cleaning or replacement of the video heads should solve the problem.

Another source of poor video performance is the tracking circuits. Don't forget to adjust the tracking (if possible) before trying to service the video circuits.

Most of the video processing in a VCR is accomplished with a few ICs. In the GE VG-7725, the main video IC is IC3001, the chroma/luma processor (**Figure 7.1**). Other video ICs are record/head amp (IC3501) shown in **Figure 7.2**, the 2FSC CCD 1H delay IC (IC3101) shown in **Figure 7.3**, and the character generator (IC3003) also shown in **Figure 7.1**. You can check the inputs and outputs of these ICs to find out if they are operating as expected. To do this you need to have an overall understanding of how the video circuits work. If you find something wrong with an IC, you need to desolder it carefully and solder in an exact replacement.

Figure 7-3. Schematic diagram showing IC3101, the 2FSC CCD 1H Delay IC.

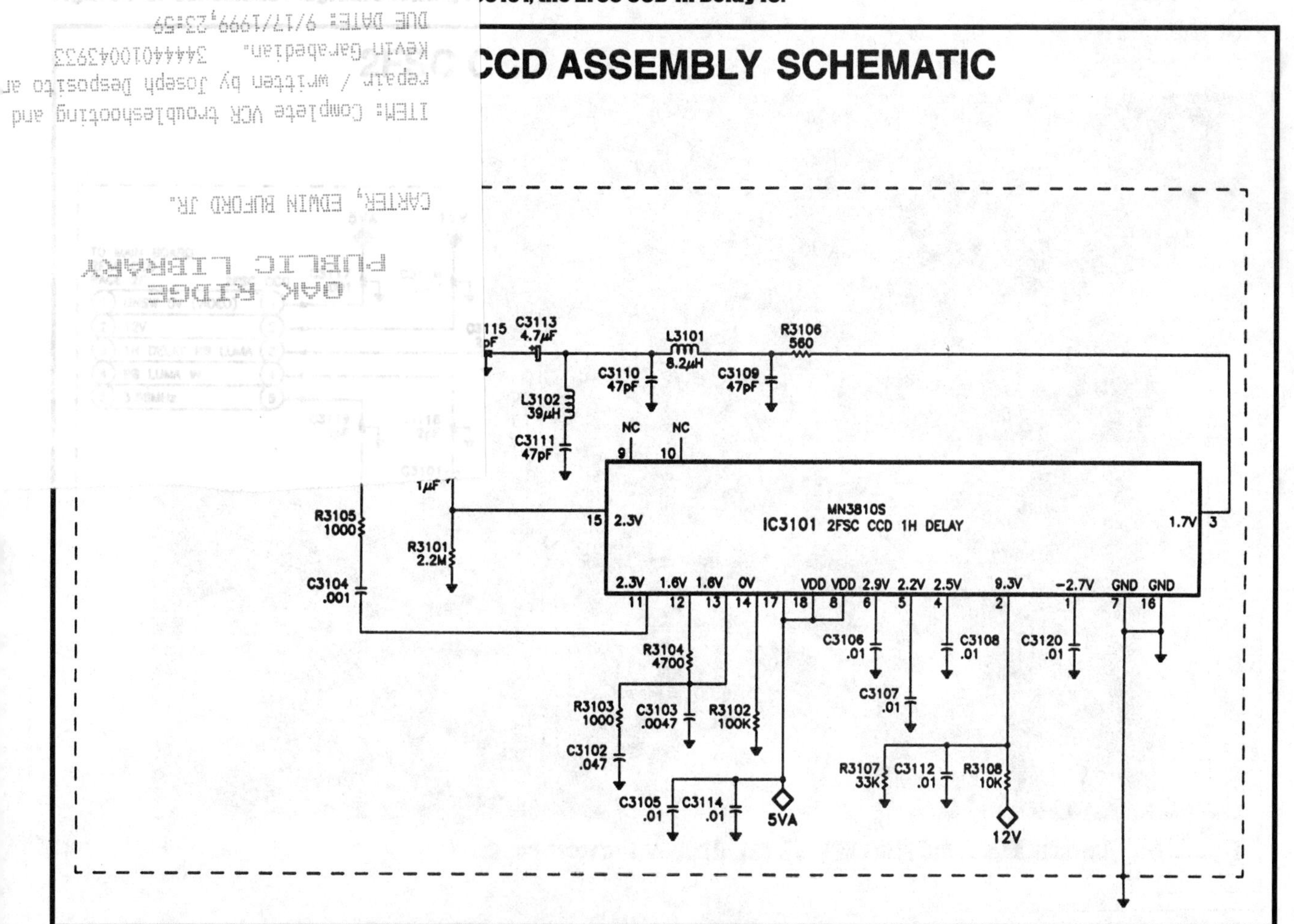

Figure 7.4 shows the area on the main board of the VCR where some of the video circuits reside. Notice the daughter board soldered to the main board. IC3001 is soldered onto this daughter board, in effect sandwiching the IC between the daughter board and main circuit board. **Figure 7.5** shows where the head amp board (containing IC3501) is located in this VCR. Notice the shielded enclosure. The positioning of the head amp board is typical of most VCRs (behind the video heads). It is particularly difficult to reach in this machine due to the main circuit board's position at the top of the chassis.

Figure 7-4. The area of the main board of the VCR where some of the video circuitry is located.

Figure 7-5. The head amp board of the VCR is located behind the video heads.

One of the servicing tools mentioned earlier in the book is a test tape. If you have a test tape and a good oscilloscope, you can perform the adjustment procedures outlined in most service manuals. This may help you to clear up the trouble with the video output.

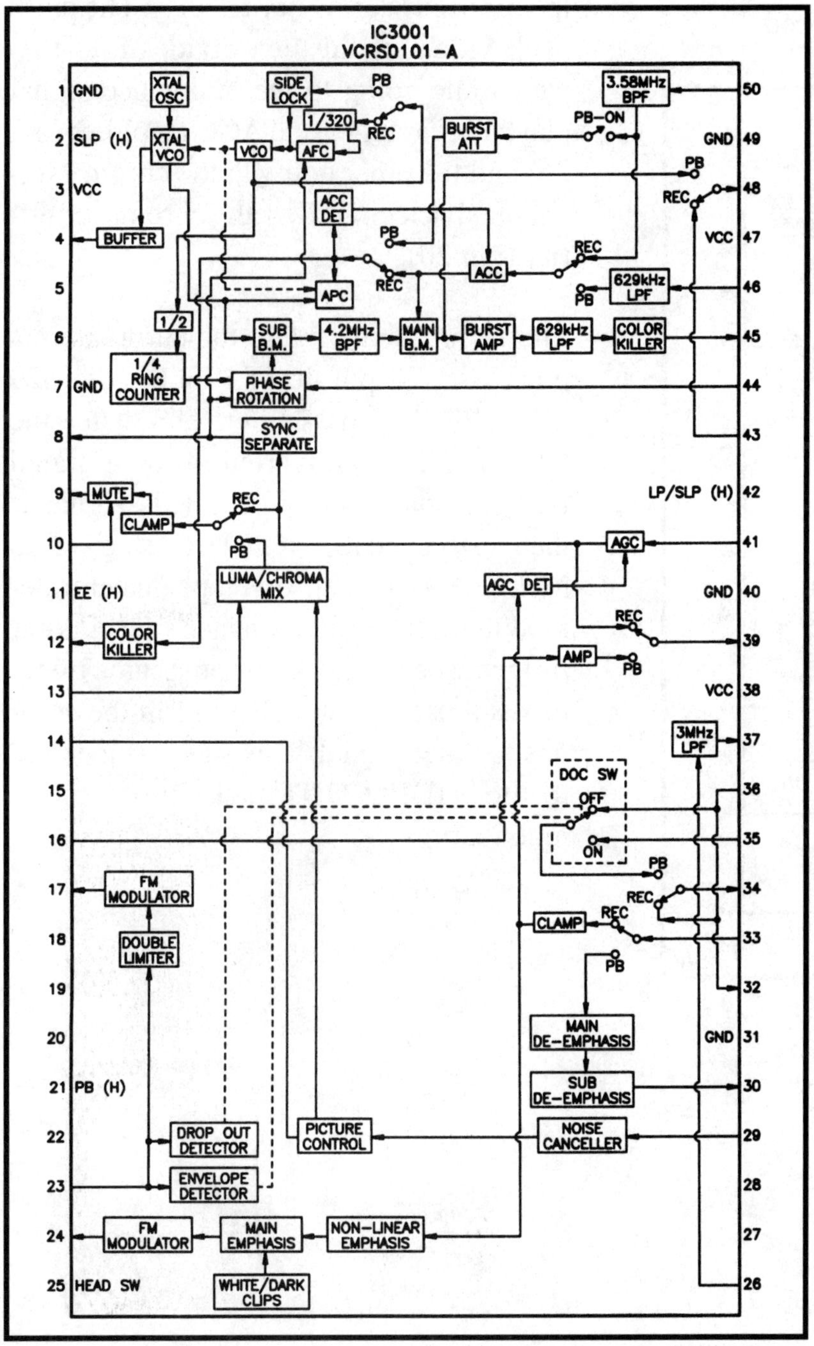

7.1 Chroma/Luminance Recording

The composite video signal cannot be recorded directly onto the video tape. It must undergo further processing because of the bandwidth limitations of the tape. Much of this processing is handled by IC3001 in the GE VCR, the chroma/luma processor. **Figure 7.6** shows IC3001 in block diagram form. Essentially, IC3001 separates the chroma and luminance portions of the video signal. The chroma (C) is the color information, and the luminance (Y) is the brightness information in the video signal.

During recording, the luma or luminance signal is frequency modulated so that its sync tip is 3.4 MHz and its white peak is 4.4 MHz. The chroma signal is downconverted from 3.58 MHz to 629 kHz and amplitude modulated. In this VCR, the frequency modulated luminance leaves the IC at pin 24. The downconverted chroma leaves the IC at pin 45. Both luminance and chroma signals are fed to IC3501, the head amplifier.

Figure 7-6. Block diagram of the Luma/Chroma Processor IC.

The head amplifier, IC3501 (**Figure 7.7**), mixes both signals together (pins 21 and 23). The resultant signal is fed to a pair of video heads, depending on the speed. If the speed is sp, the signal goes to the sp recording heads. If the speed is lp or slp, the signal goes to lp/slp heads. On any speed, the VCR uses only one pair of heads. These heads are optimized for that speed.

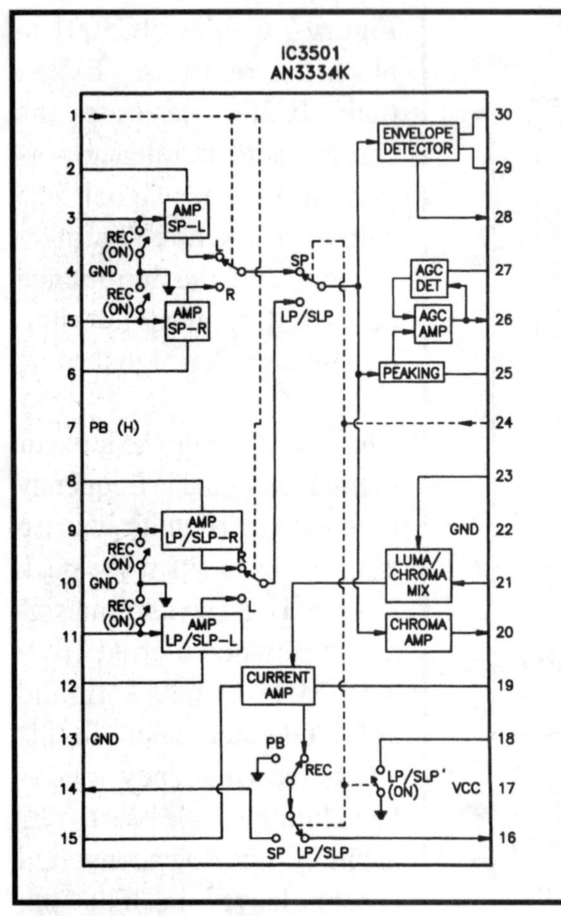

Figure 7-7. Block diagram of the Record/Head Amp IC.

7.2 Chroma/Luminance Playback

During playback, the video signal on the tape is picked up by the video head. This weak signal is amplified by IC3501 and passed through the head switching circuits. These circuits switch between the sp and lp/slp speeds during normal play. The playback video signal then divides into two paths, one going to the luminance signal playback circuit via the AGC amplifier (pin 26) and the other going to the chroma signal playback circuit via the chroma amplifier (pin 20).

These signals are fed to the chroma/luma processor. At pin 46 of IC3001, the 629 kHz LPF (low pass filter) eliminates the 629 kHz downconverted chroma signal from the playback signal. This signal is then restored to its original frequency (3.58 MHz). At pin 23 an envelope detector demodulates the FM luminance signal. After further processing, the chroma/luma mixer mixes these two signals to form the composite video signal. This signal is then fed to the VIDEO OUT jack and to the rf modulator.

Chapter 8

Troubleshooting Audio Circuits

The audio signal comes from the tuner/demodulator. In the GE VG-7725, the audio signal can be found at pin 28 of IC701 (refer to **Figure 6.2** in Chapter 6). This signal is fed to pin 11 of IC4001 (**Figure 8.1**), the audio switching IC, also called the audio recording and playback processor. The block diagram of the chip is shown in **Figure 8.2**. Its position on the printed circuit board is shown in **Figure 8.3**. You'll notice that this IC is very easy to get to. The way this VCR is designed, the main board sits foil side up as you look into the machine. The audio IC is easily accessible from this side of the board. If you swing the board up, the IC is also in a good spot to work on. The IC has only 24 pins so it's easier than most of the other major ICs to desolder and replace.

IC4001 connects directly to the audio head. This head is basically a universal head; in other words, it is used for recording and playback, just like the video head.

8.1 Audio Recording

Once the audio signal enters IC4001, the recording circuits of the IC take care of processing the signal. Initially, the audio signal gets emphasized by a circuit for the high frequencies. Then the signal is mixed with the frequency from the local oscillator, which normally has a high frequency about 50 kHz.

This signal biases the tape so that the recording is in the linear region of the magnetic hysteresis curve. The audio signal is recorded on a small, very narrow track at the top of the video tape. When recording sound, a problem can arise if the bias oscillator for some reason stops working; for example, the transistor goes bad in the oscillator or the coil goes bad. Then the picture does not get erased because part of the bias signal goes to the full erase head to erase the signal from the previous recording. The VCR may record the new picture, but when you move the tracking adjustment to one side you get one recording and when you move it to the other side you get a completely different program. It's like having two recordings on the same stretch of tape. On the recording side, the bias oscillator is very important. It always has to be checked. The bias oscillator is Q4002 in **Figure 8.1**.

Most VCRs have a test point to check the bias level on the erase head and the recording head. It's very important with the recording head to check the bias current. This affects the level of the recording, which is responsible for the audio quality. If the bias level is too high, some of the high frequencies may be erased. If it's too low, distortion will occur. Basically, this is done with the bias level control, which is R4022 in the GE VCR. Some VCRs do not provide an adjustment for the bias level. If this is the case, you can turn the coil of the oscillator to adjust the bias level.

Keep in mind that the audio head has to be aligned properly in order to record properly. In other words, the height of the audio head is very important as well as the azimuth. This is so that a tape recorded in one machine can be played successfully in another machine. In general, you should not expect any problems with the recording process unless somebody has tried to adjust certain parts.

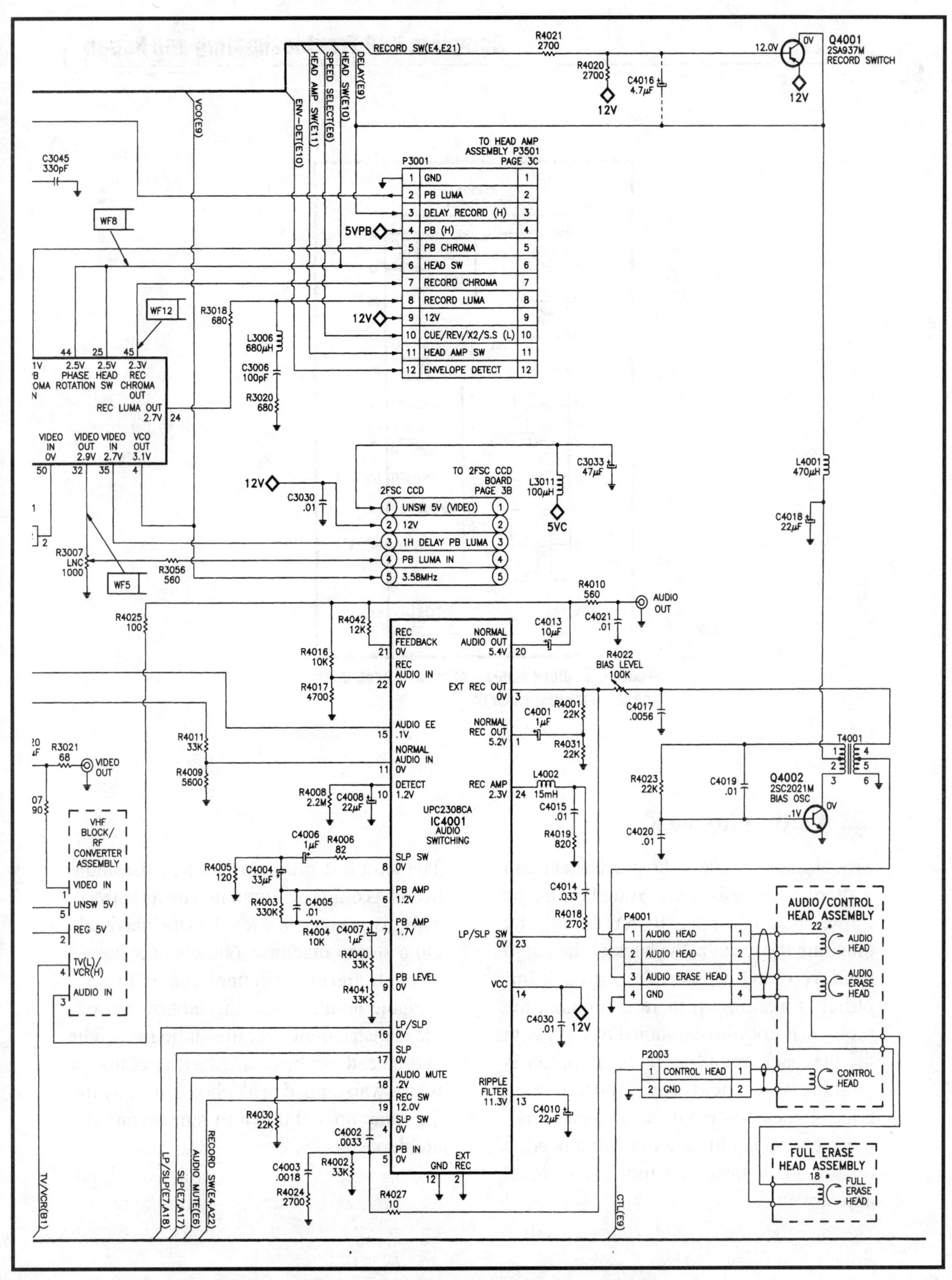

Figure 8-1. Schematic diagram showing IC4001, the audio recording and playback processor.

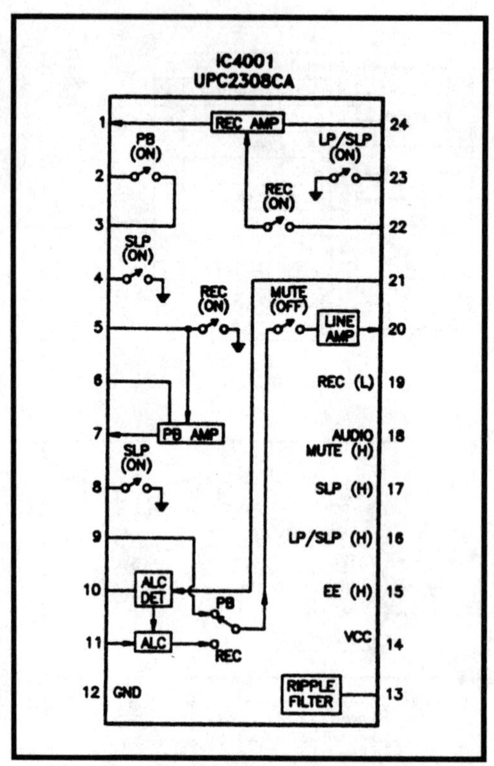

Figure 8-2. Block diagram of the audio recording and playback processor IC.

8.2 Audio Playback

Troubleshooting the audio playback circuit is pretty straightforward. You place a tape in the VCR and press PLAY. Simply by touching the wires at the back of the audio head, you can find out if the playback amplifier is working. If there is no hum, the problem is not with the audio head, it's with the playback amplifier. If the amplifier is working, you should hear a loud humming noise. If the amplifier does produce sound, the problem could be a defective head, a dirty head, or somebody touched the level adjustment.

The simplest problem is when the audio head has oxide accumulated on its surface. This gives you a low level on the playback. On a stereo machine, one channel comes through and one channel comes in on a variable level, depending on how the oxide is distributed over the audio head. The audio head can be cleaned with a cotton or foam swab dipped in alcohol. Carefully rub the swab up and down to remove any oxide deposits.

Figure 8-3. The audio recording and playback processor IC on the VCR's main circuit board.

8.3 Mechanical Adjustments

To improve audio quality, you may have to adjust the tension of the tape. Audio recording requires a good mechanical contact between the audio head and the tape. Other adjustments you can perform are for the height and azimuth of the audio head.

To adjust the height of audio head, you need a gauge normally sold by the manufacturer of the VCR for this purpose. But you can do it without the gauge if you develop certain skills. Use a tape recorded on a VCR that is in good condition. Insert the tape into the VCR under repair and press PLAY. The audio head is mounted on a spring loaded shaft. By turning the spring up and down, you can adjust the height of the head. Connect a DMM set to the volts setting to the AUDIO OUTPUT RCA jack at the back of the VCR. By moving the head up and down you can determine the maximum level of the sound. This is the best position to leave the head.

The azimuth is adjusted the same way. The only difference is that when you turn the screw for the azimuth, the head tilts left and right. You have to check it with a tape recorded at a high frequency, in the range of 7 to 10 kHz. When you adjust the azimuth, the maximum voltage level will occur at the highest frequency. This is the right azimuth. You can use a factory tape or one you made yourself on a good VCR.

The signal from the playback head (also is the recording head) goes to the pre-amplifier, which normally has an adjustment for the left and the right channels on a stereo machine. On a monaural machine, there is an adjustment for a playback level. This level should be comparable to the level of the audio from the tuner.

Some machines have provisions for adjustment, some machines don't (playback level is set at the factory). Because the audio head is a combination audio head and control head, which reads the speed, there is a small PC board that sits behind the audio head. Sometimes a solder joint cracks and breaks the connection. If this happens, you may hear a humming noise or the VCR may not record at all. Also, playback will be very low with a humming noise. The connections of the audio head should be checked out anytime you have any audio problems. And, if needed, the connection should be resoldered. On some VCRs, the audio connector may go bad. One solution is to cut the wires from the connector and solder them directly to the head to avoid future problems.

8.4 Stereo Hi-Fi Circuits

Audio circuits are more difficult to troubleshoot on a stereo VCR. There are two channels, left and right. If you have a discrepancy between the two channels, for instance, one channel is higher than the other, then you have to find the control marked playback or bias level and adjust it. You need to have a certain quality test tape to do this. First, you do the playback adjustment and then you do the recording adjustment. You have to have some reference point when you start doing the playback adjustment. Make sure the left and right channels are equal. Then you have to make sure the system records the same way. Recording can be adjusted if the VCR has an adjustment for recording level.

Chapter 9

Troubleshooting Special Effects Circuits

9.1 Search, Fast Forward and Rewind

When a VCR is playing and you press fast forward, this effect is called search fast forward. You can see the picture while the machine is fast forwarding. This condition depends on a few components, for instance, the mode switch (**Figure 9.1**). If it is not properly adjusted or aligned, or if it has dirty contacts, the VCR may stop playing altogether when you press fast forward. Also, mechanically, if the idler arm assembly (**Figure 9.2**) or a rubber bumper is worn out, the tape will not speed up.

The clutch assembly (**Figure 9.3**) is very sensitive to the search rewind special effect. In this mode, the idler assembly pulls the tape backward and the pinch roller with the capstan releases the tape.

Special effects like fast forward and rewind can be pretty easy to check. First of all, you have to put in a good tape recorded on a slow (EP or SLP) speed. Most machines with two heads may have a pretty good picture on EP. Then, while the machine is playing press rewind and check if the machine shows the picture, even if there are a few thin lines across the screen. But, basically, the picture has to be visible. Then you know the machine is working in that mode. Then, you can check fast forward while the machine is playing. The machine may produce a few lines, but the rest of the picture should play clear and fast moving. If mechanically the machine is aligned properly, it will work in this mode.

The reverse gear on some models, such as early Panasonic VCRs, engages only on a search rewind. This gear moves the tape backward. It is activated by the loading

Figure 9-1. The mode switch must be working properly for the VCR to perform operations such as search, fast forward and rewind.

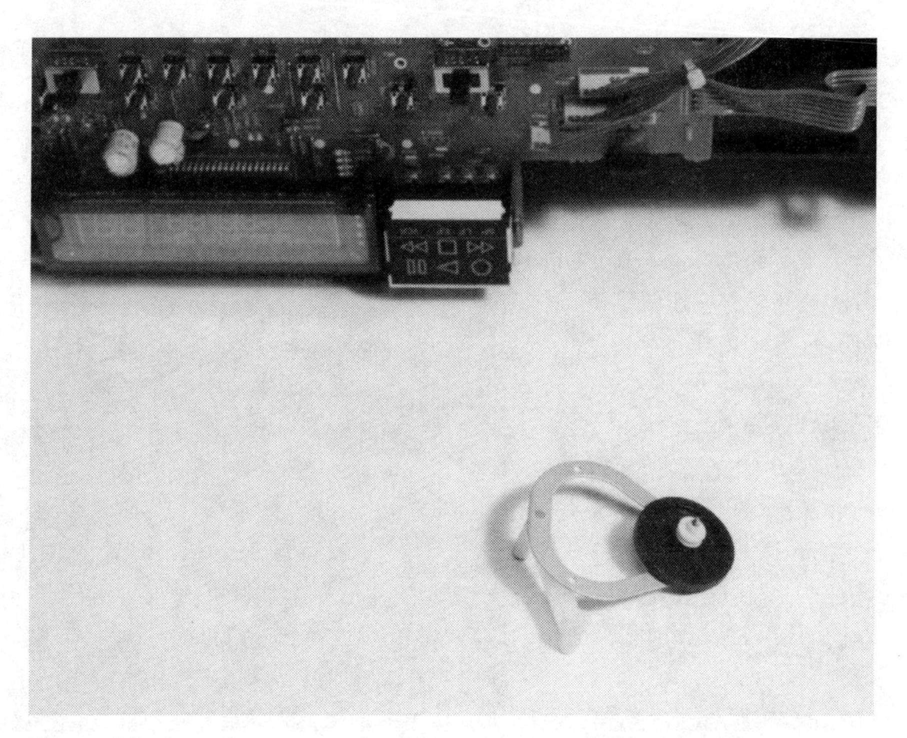

Figure 9-2. If the idler assembly tire is worn out the VCR will not speed up.

assembly underneath the machine. So when you press the REW button, the loading motor engages that gear and it turns the tape. It is very sensitive to the condition of the loading belt. If the loading belt is a little worn out the VCR will not go into search rewind. It will shut off. Also, these VCRs are very sensitive to the clutch and idler arm assembly. It is a very complicated system. Nowadays, no VCRs use this system. Normally, the idler assembly will go back to the left wheel and will start turning the wheel backward. Meanwhile, the capstan shaft with the pinch roller turning backward will release the tape. In other words, the capstan pulls the tape out of the cassette on one side and on the other side the take-up reel collects the tape inside of the cassette. Also, all the tension brakes are released.

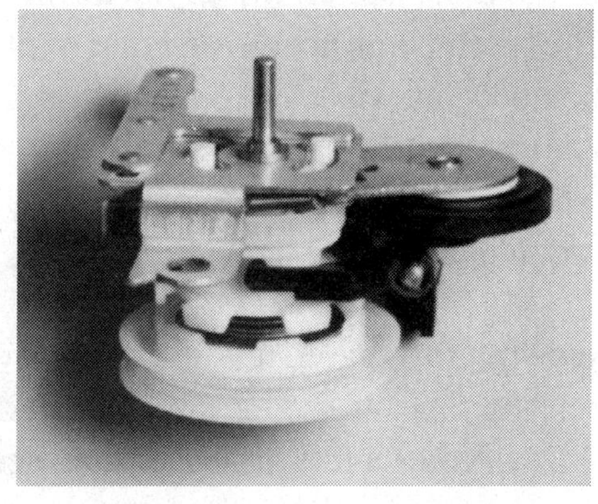

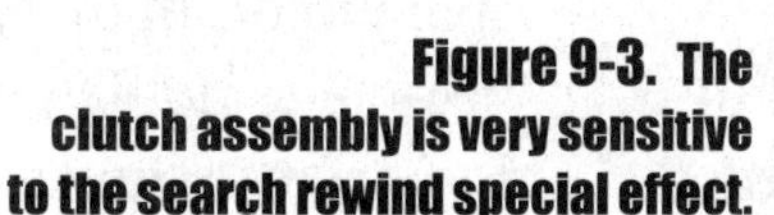

Figure 9-3. The clutch assembly is very sensitive to the search rewind special effect.

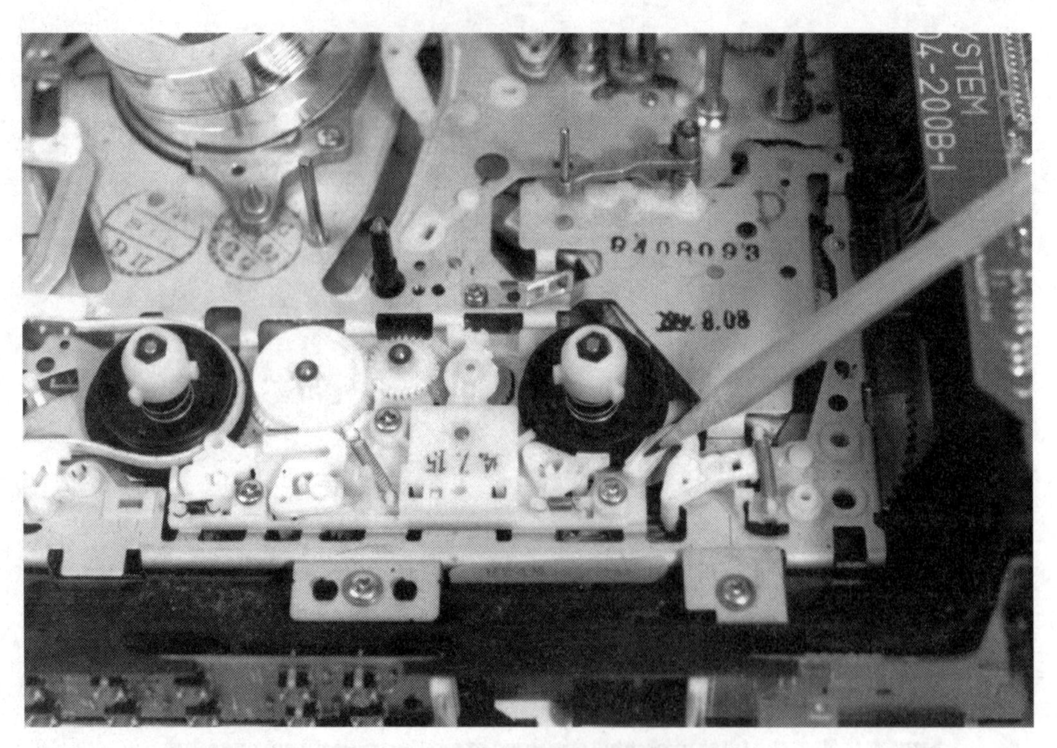

Figure 9-4. A problem will occur if the VCR does not release the brakes.

A problem occurs if the VCR does not release the brakes (**Figure 9.4**) or if there is heavy tension in the tape. This has to be checked. Many times when you press search rewind, the machine starts moving the tape and then it stops because the back tension is very heavy. In other words, the tape has to have a very light tension for search rewind to work properly. To fix this, you may have to replace the capstan shaft, the idler arm assembly, or in some cases even the pinch roller. If the pinch roller is a little worn out, it may work fine in playback mode, but when you go into the search rewind mode, the tape gets pulled up the capstan shaft or pulled down, and the tape gets damaged by the tape guide.

When you do a search rewind, you have to watch the tape. If the tape is damaged in any way, you have a problem that needs repair.

9.2 Slow Motion

Slow motion is different from fast forward and fast rewind because the machine actually enters a different mode. The microprocessor signals the capstan motor to slow down the speed. Some VCRs have a provision for how much the machine can slow it's speed. Some machines have a fine adjustment for the slow motion speed. If the machine plays normally, though, the slow motion will work pretty well (except if it's a two head machine and you try to do slow motion in the SP speed).

A problem with slow motion is rare. A problem with slow motion means a problem with the microprocessor or the capstan motor. In this case, you will not even be able to do playback. If everything else is working, slow motion will work.

Chapter 10

Troubleshooting Sensors and Switches

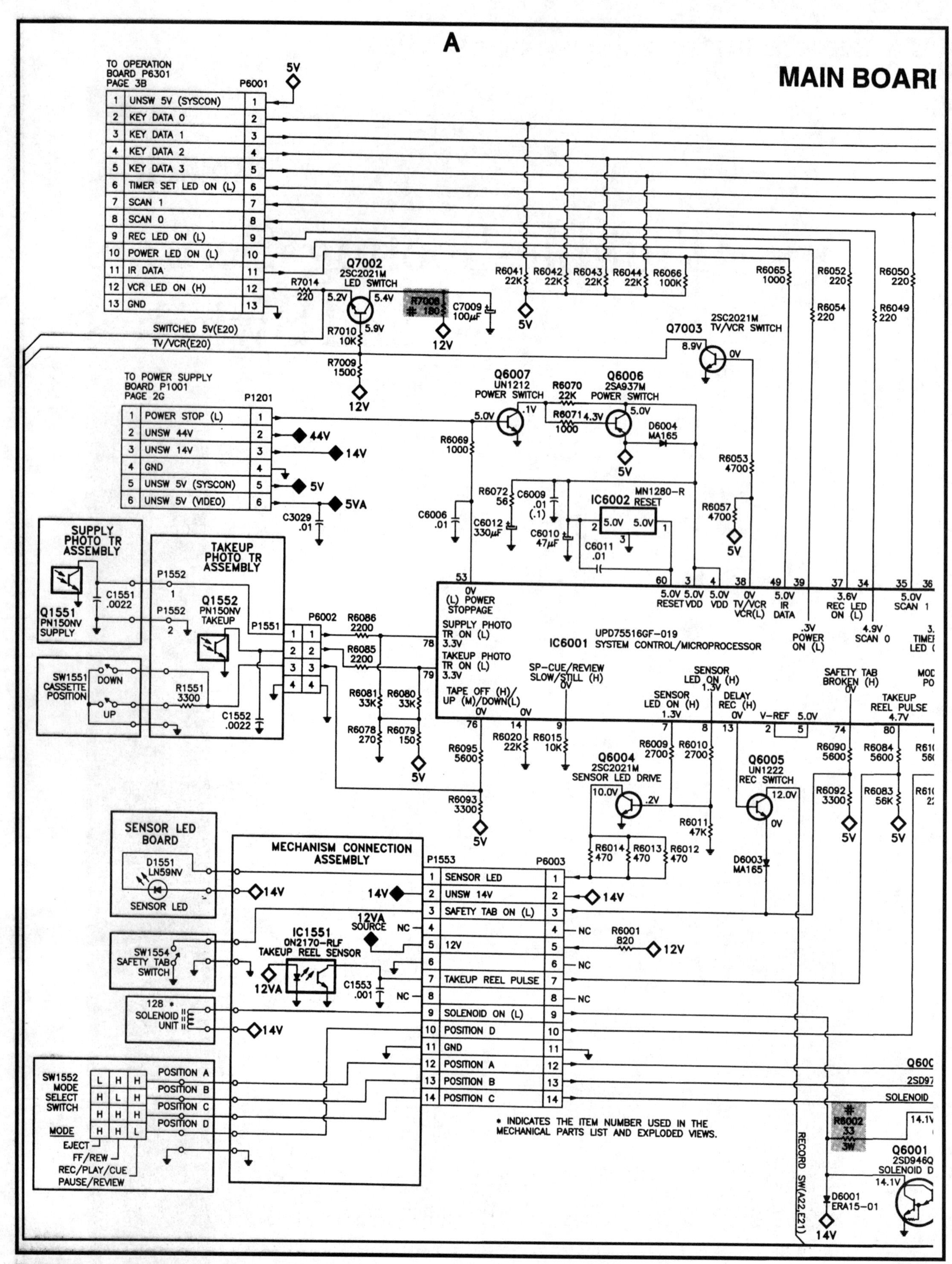

Figure 10-1. A schematic diagram showing the sensors that go to the microprocessor.

Sensors and switches provide information to the microprocessor about the current status of the VCR. If a sensor or switch fails, the result is havoc in the VCR. The microprocessor turns the VCR off, when it should be on, it fast forwards a tape when it should be rewinding, and so forth. Remember, a VCR is a very sophisticated electro-mechancial system, dependent on the precise interaction of hundreds of parts. When one small part fails, often the whole system fails.

10.1 Checking Sensors

The microprocessor takes inputs from many sensors. **Figure 10.1** shows the sensors of the GE VG-7720 VCR. Note that they all act as inputs to the microprocessor. VCRs use all different kinds of sensors. This particular model uses a photointerrupter and two phototransistors. The phototransistors receive light from a separate LED, while the photointerrupter is a self-contained unit, both sending and receiving light.

A tape-end sensor is often found on the cassette housing but also may be mounted on the main circuit board in some VCRs. When the tape reaches the end of its run, there is a clear portion of tape known as a transparent leader. Light from an LED passes through the tape, a phototransistor senses the light, the microprocessor receives the signal and responds according to the current mode of operation. If in the PLAY mode, the operation is switched to rewind. A tape-start sensor stops the tape after it is finished rewinding. This is done through a transparent leader at the beginning of the tape.

Figure 10.2 shows the chassis of Mitsubishi HS-U260. The sensors are partially hidden by the chassis.

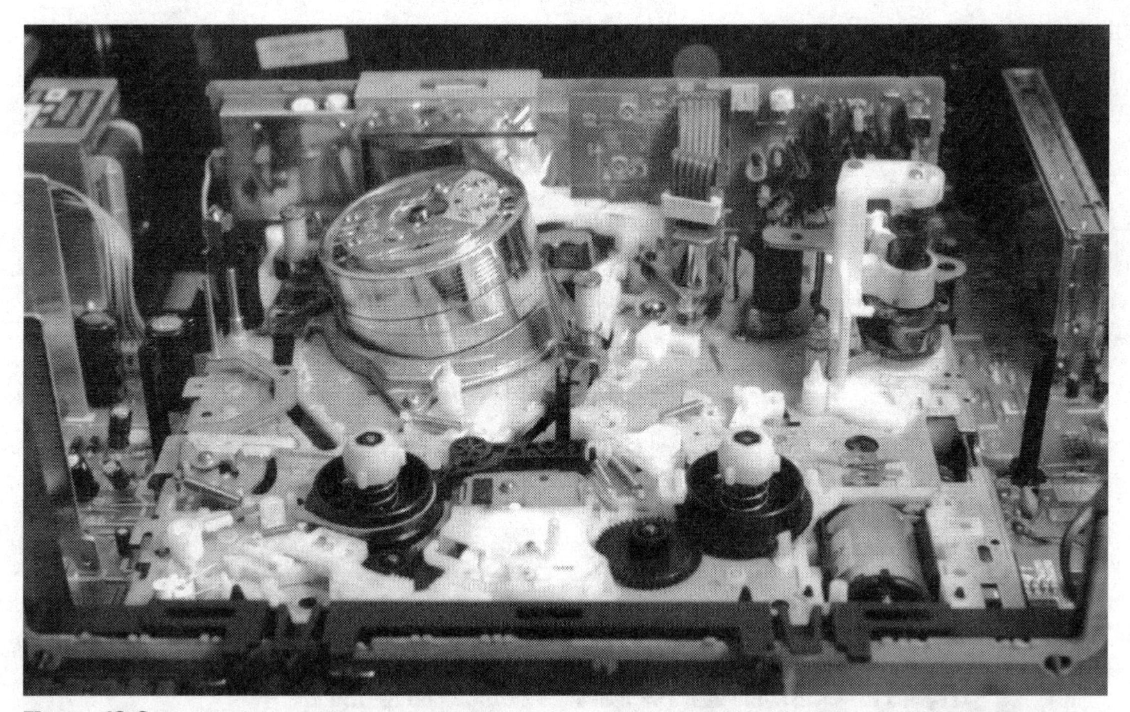

Figure 10-2. It's hard to see the sensors when looking at the chassis of the Mitsubishi HS-U260.

Figure 10-3. With the chassis removed, it's easy to see the "triple tower" sensors attached to the main circuit board of the Mitsubishi HS-U260.

Figure 10.3 shows the same VCR with the chassis removed. Now, it is very easy to see the "triple tower" sensors connected to the main PC board. The center tower contains the LED, while the two end towers contain the phototransistors. The phototran-sistor on the left side is used as a tape-end sensor, the one on the right is a tape-start sensor. Other VCRs have the LED mounted on a post in the middle of the chassis, while the phototransistors are mounted on the left and right sides of the cassette housing (**Figure 10.4**).

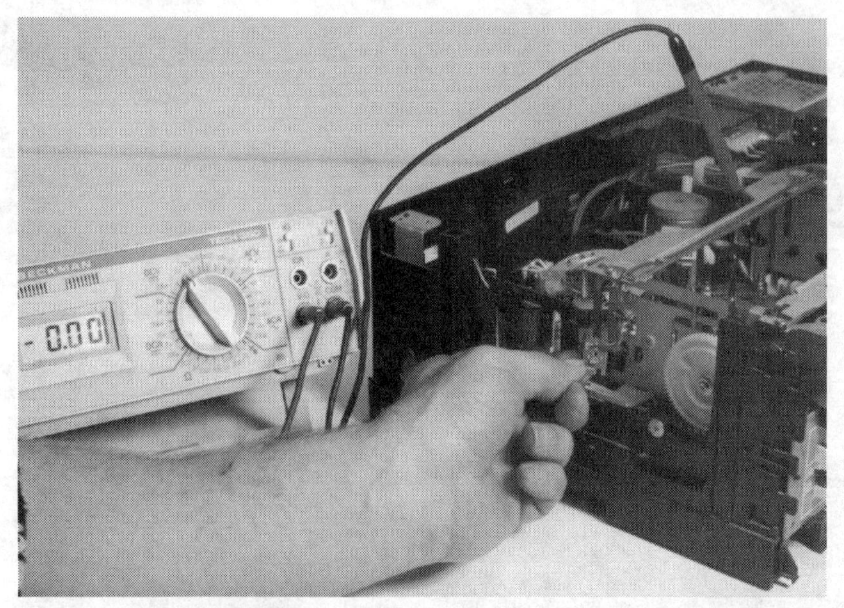

Figure 10-4. On some VCR's sensors are mounted on the left and right sides of the cassette housing.

Figure 10-5. Underneath the take-up or supply reels is a pie shaped pattern with alternate dark and shiny wedges.

Photointerrupters are often found underneath the take-up and supply reels and are used as a tape counter. A photointerrupter is a 4-pin device with a light emitting diode on one side and a phototransistor on the other. Underneath the take-up or supply reel is a pie-shaped pattern, which alternates between shiny and dark wedges (**Figure 10.5**). The light from the LED shines on the pattern and reflects back to the phototransistor. As the reel turns, a pattern of pulses is created, which is sent to the microprocessor. The count is displayed on the front panel of the VCR or on screen. If this sensor is not working the machine will shut down.

Figure 10-6. A photo interrupter used as a tape-in sensor.

Some VCRs use a different type of photointerrupter, such as a tape-in sensor. This type of photointerrupter is shown in **Figure 10.6**. Physically, the device contains two upright blocks with a gap in between. On one side is the LED and on the other side is the phototransistor. A tab on the

Figure 10-7. A tab on the cassette housing sits in the gap of the photo interrupter.

cassette housing (**Figure 10.7**) sits in the gap, preventing light from hitting the phototransistor. When a tape is loaded, it pushes the tab out of the gap. The light from the LED hits the phototransistor, sending a signal to the microprocessor that the tape is in the machine.

The dew sensor is another popular sensor. Some machines have an indicator on the front panel for the dew sensor, some don't. The dew sensor changes its resistance depending on the moisture in the VCR. Moisture can cause the tape to stick to the highly polished video head. This is why most machines have this provision. If there is a humidity buildup inside the machine the microprocessor ensures that nothing will work until everything dries. You can speed up the drying process with a hair dryer.

You can check the status of a sensor with a DMM set to measure DC voltage. The voltage at a sensor varies from high to low, depending on whether the sensor is on or off. Voltages vary. Some sensors may measure up to 10 volts when on, others may measure only 2 volts. When taking the measurement, the primary concern is whether or not the voltage switches from high to low when you activate or deactivate the sensor. With an LED/ phototransistor combination, you can place your finger or an object over the LED or the phototransistor itself. With a self-contained device such as some photointerrupters, you have to use the reflective surface to change the voltage. In other words, you would have to turn a take-up reel by hand to get the sensor to change voltages. With other kinds of photointerrupters, such as the one with the two upright blocks, you can insert an opaque object between the two blocks to test the sensor. For a dew sensor, you can wet your finger, place it right on the sensor, and take the measurement.

10.2 Checking Switches

Of the many switches that provide information to the microprocessor, the mode switch is the most complex. In the GE 7720, the mode switch is a 4-position switch (**Figure 10.1**). It connects directly to the microprocessor. Thus, before the microprocessor resets itself, the mode switch has to be in good working condition. If the mode switch is worn out or dirty and doesn't make the proper connections, chances are the VCR will shut off prematurely. For example, if you press the PLAY button, the VCR will start performing the function and then will shut down. It may even seem as though there is a problem with the microprocessor. But as we have said, usually if the microprocessor fails, nothing at all will work.

The mode switch is the only switch in the VCR that needs to be aligned correctly in order to work. In most VCRs, the mode switch looks like a toothed gear attached to a PC board (**Figure 10.8**). On the gear, there is a dot that must be aligned with an arrowhead or dot engraved on the plastic base of the switch or on another gear. This alignment also depends on the position of the tape in the machine. Usually, the tape has to be out of the machine when the dots are aligned. The biggest reason the mode switch loses its alignment is that the owner of the VCR forces a tape into the machine.

Figure 10-8. The mode switch is attached to the PC board.

If the mode switch is dirty, it can be cleaned by spraying WD-40 between the wheel and base of the switch. You should spray a few times until you are sure the contacts are clean.

Figure 10-9. The safety tab switch.

Another popular switch in a VCR is the safety tab switch. **Figure 10.9** shows the position of a safety tab switch in a GE Model 1VCR5011X. This switch prevents users from recording on tapes whose safety tab is broken (usually prerecorded tapes). If this switch gets dirty, it will affect the VCRs ability to record on blank tapes. This switch can be cleaned by spraying it with WD-40 or by scraping dirt or oxide deposits off the switch with a small screwdriver.

On some VCRs, such as the GE 7720, the position of the cassette is monitored with a mechanical switch rather than a sensor. Typically, there are two switches involved, one for cassette down and one for cassette up. If this switch becomes dirty, it can be cleaned in the same way as the safety tab switch.

Chapter 11

Troubleshooting Microprocessor Control Circuits

Big Brother has never been so alive and well as in modern VCRs. The microprocessor watches over all the circuitry. If it finds something wrong, it shuts down the VCR after a few seconds. And if the microprocessor itself fails, you cannot even turn the power on. In older models, the power would come on but nothing would work, you could not put the tape in. You might have to replace a microprocessor in these older machines, but it is very rare to have to replace one in models built within the past ten years.

Not only have manufacturers improved the technology, they also place a fuse in the 5 V line that powers the microprocessor (**Figure 11.1**). This fuse looks just like a small transistor. The microprocessor is one of the best protected circuits in the VCR. If it does fail, usually the failure is caused by a strong electrical surge or an act of God.

The GE VG7720 VCR employs a UPD75516GF-019 microprocessor, IC6001. **Figure 11.2** shows the schematic diagram of the system control section of the VCR. **Figure 11.3** shows the location of the microprocessor on the main printed circuit board of the GE VG-7725 VCR. This model uses a MN188166VLC3 microprocessor. Note that the microprocessor is soldered to the foil side of the board in this machine. Also, the microprocessor is a surface mount package. Replacing this IC would require excellent soldering skills as well as experience with surface mount technology. On the other hand, the positioning of the microprocessor makes it easy to take measurements with a DMM, oscilloscope, or logic probe.

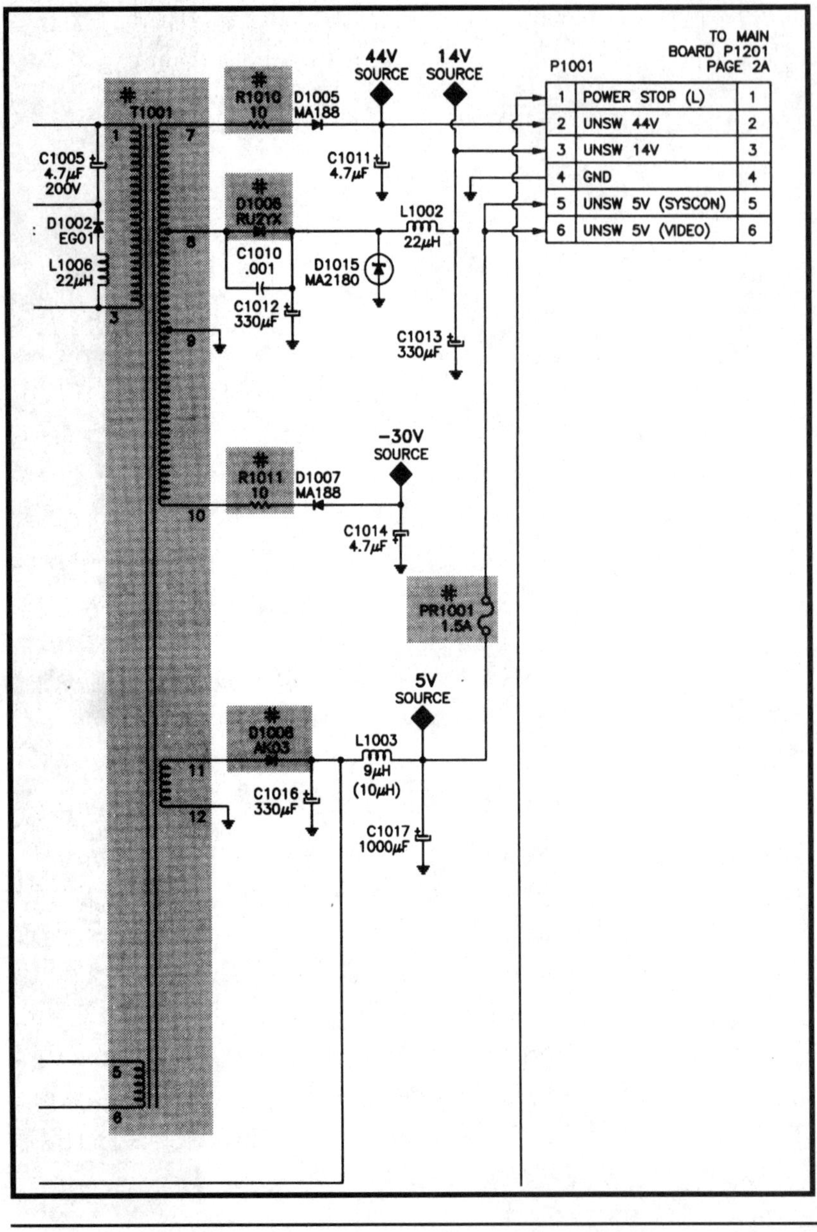

Figure 11-1.
The power supply contains a protective fuse in the 5V line that powers the microprocessor.

Figure 11-3. Location of the microprocessor on the main circuit board.

1.11 Checking the Microprocessor

You may want to make some checks of the microprocessor, just to better understand how it operates. Some interesting points to check are the supply voltage, clock input, remote control input, sensor and switch inputs, and reset. In the GE 7720, the 5 V supply voltage, VDD, is at pins 3 and 4. The crystal is connected between pins 58 (X1) and 59 (X2) and runs at 4.19 MHz. Remote control signals come in on pin 49, IR DATA. Sensor inputs come in at pins 7, 8, 74, 76, 78, 79, and 80. Mode switch inputs come in on pins 65 through 68. Checking how these signals react as you perform certain functions, such as loading a tape, provides a good insight into the microprocessor's role in the VCR.

To check how an output responds, select one control. Let's say select PLAY mode. Pressing PLAY on the remote sends a signal to the IR DATA input. Pressing PLAY on the VCR control pad sends a signal to pin 30, KEY DATA 1. **Figure 11.4** shows the keypad schematic of the GE 7720. Next, locate the output of the microprocessor corresponding to the PLAY command. For the 7720 VCR, this is pin 63, PLAYBACK ON. Put the voltmeter there and press the PLAY button. You'll see the voltage change from low to high or high to low, whichever is the case for that output.

You can also try this experiment on any VCR you are working on. With power removed from the VCR, desolder the pin of the microprocessor connected to 5 V. The easiest way to locate this pin is to check the schematic diagram of the VCR. If you don't have the diagram, you can find this pin by inspecting the main PC board of the VCR or taking measurements with a DMM.

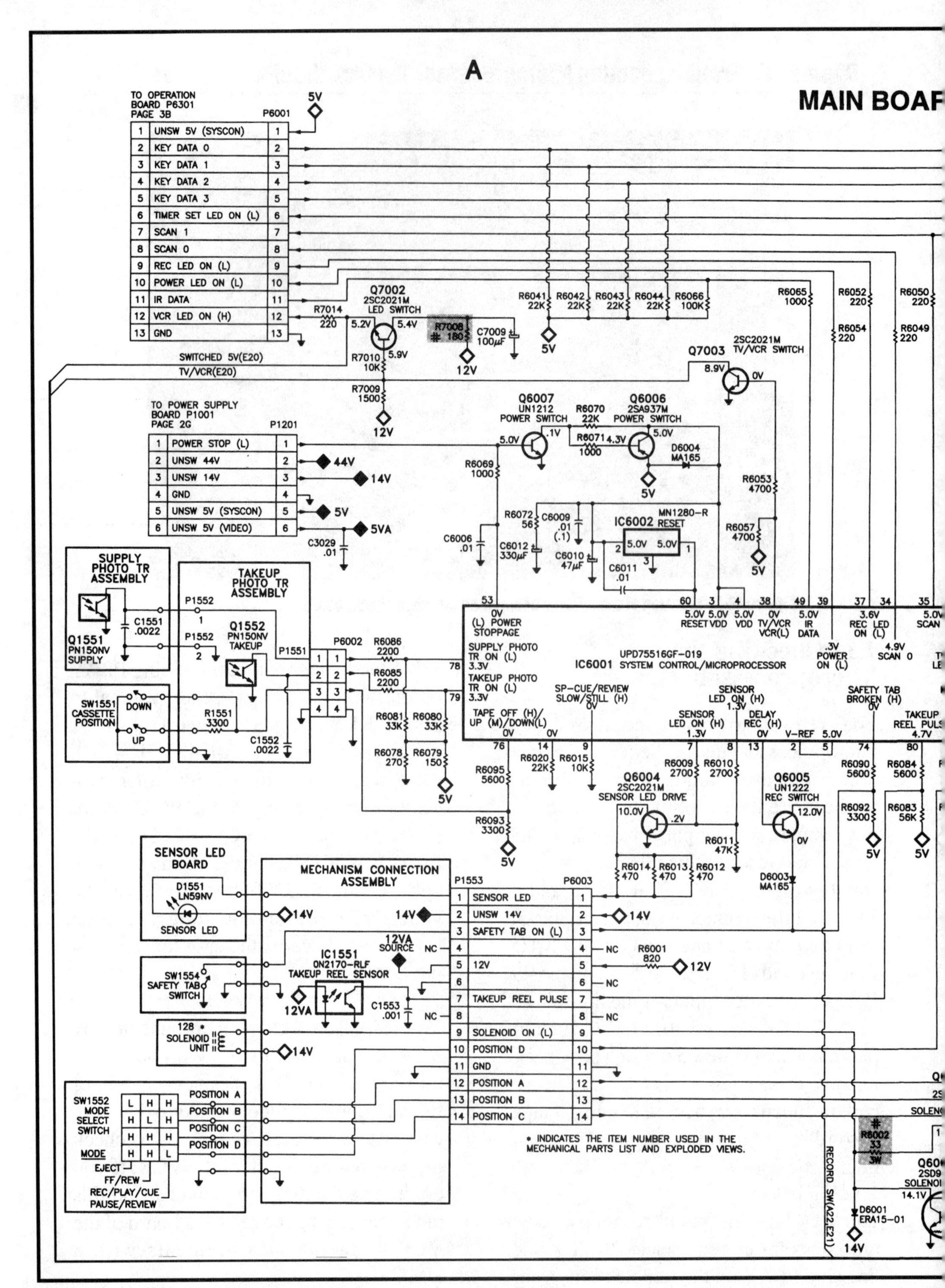

Figure 11-2. The schematic diagram of the system control section of the VCR.

B

HEMATIC

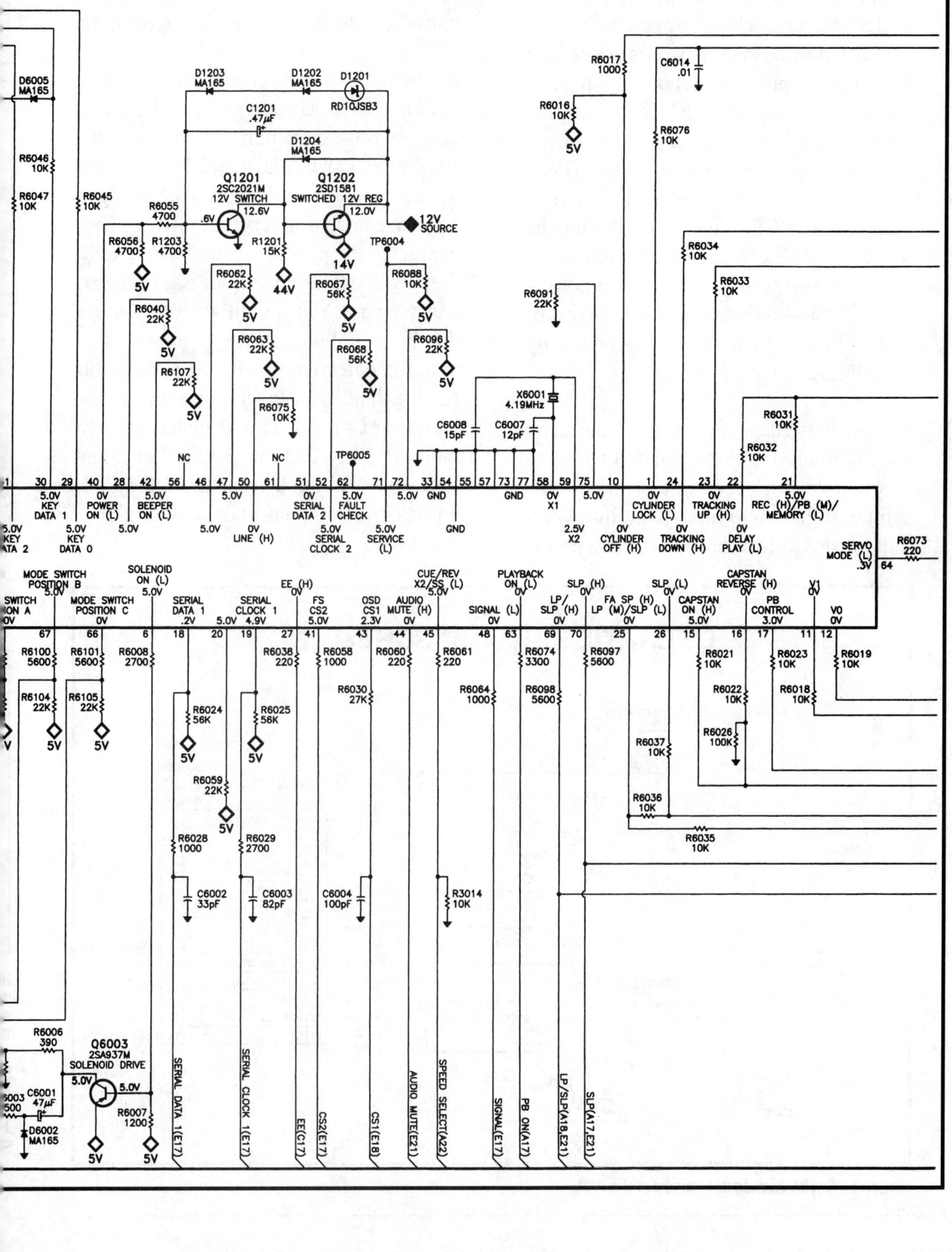
D6005
MA165
R6046
10K
R6047
10K
R6045
10K
D1203
MA165
D1202
MA165
D1201
RD10JSB3
C1201
.47µF
D1204
MA165
Q1201
2SC2021M
12V SWITCH
12.6V
Q1202
2SD1581
SWITCHED 12V REG
12.0V
12V
SOURCE
R6055
4700
.6V
R6056
4700
R1203
4700
R1201
15K
14V
44V
TP6004
R6062
22K
R6067
56K
R6088
10K
R6040
22K
R6063
22K
R6068
56K
R6096
22K
R6107
22K
R6075
10K
TP6005
NC
R6017
1000
C6014
.01
R6016
10K
R6076
10K
R6034
10K
R6033
10K
R6091
22K
X6001
4.19MHz
C6008
15pF
C6007
12pF
R6031
10K
R6032
10K
KEY DATA 1
KEY DATA 0
KEY DATA 2
POWER ON (L)
BEEPER ON (L)
LINE (H)
SERIAL DATA 2
SERIAL CLOCK 2
FAULT CHECK
SERVICE (L)
GND
X1
X2
CYLINDER LOCK (L)
CYLINDER OFF (H)
TRACKING UP (H)
TRACKING DOWN (H)
DELAY PLAY (L)
REC (H)/PB (M)/ MEMORY (L)
SERVO MODE (L)
R6073
220
MODE SWITCH POSITION B
SWITCH POSITION A
MODE SWITCH POSITION C
SOLENOID ON (L)
SERIAL DATA 1
SERIAL CLOCK 1
EE (H)
FS CS2
OSD CS1
AUDIO MUTE (H)
CUE/REV X2/SS (L)
SIGNAL (L)
PLAYBACK ON (L)
LP/ SLP (H)
SLP (H)
FA SP (H) LP (M)/SLP (L)
SLP (L)
CAPSTAN ON (H)
CAPSTAN REVERSE (H)
PB CONTROL
V1
V0
R6100
5600
R6101
5600
R6008
2700
R6104
22K
R6105
22K
R6024
56K
R6025
56K
R6038
220
R6058
1000
R6060
220
R6061
220
R6030
27K
R6064
1000
R6074
3300
R6098
5600
R6097
5600
R6021
10K
R6022
10K
R6026
100K
R6023
10K
R6018
10K
R6019
10K
R6037
10K
R6036
10K
R6035
10K
R6059
22K
R6028
1000
R6029
2700
C6002
33pF
C6003
82pF
C6004
100pF
R3014
10K
R6006
390
Q6003
2SA937M
SOLENOID DRIVE
C6001
47µF
D6002
MA165
R6007
1200
SERIAL DATA 1(E17)
SERIAL CLOCK 1(E17)
EE(C17)
CS2(E17)
CS1(E18)
AUDIO MUTE(E21)
SPEED SELECT(A22)
SIGNAL(E17)
PB ON(A17)
LP/SLP(A18,E21)
SLP(A17,E21)

Make sure the pin is completely free of solder and not touching any part of the hole in which it rests. In other words, make sure that the microprocessor is not receiving any power. Like any other IC, the microprocessor needs power, usually 5.0 V, to operate. Taking power away simulates a dead microprocessor. After you have done this, plug in the VCR. You'll notice that the VCR is completely dead. Nothing is working, not even the display. This symptom—a dead VCR—indicates a failed microprocessor. Finish up this little experiment by resoldering the 5 V pin.

It's true that a dead VCR can be caused by a faulty microprocessor. But don't jump to this conclusion every time you see this symptom. It's more likely that this symptom is caused by a problem in the power supply. As mentioned, microprocessors rarely fail. Power supplies fail all the time.

Removing a microprocessor takes a lot of time and skill. Usually, you will have to desolder 60 or more pins, depending on the VCR model. You'll also need a low-wattage iron with a fine point to do the job right. You have to do this very carefully, without damaging the printed circuit board. Additionally, the microprocessor is a fairly expensive part. You do not want to remove this piece unless you are absolutely certain it is defective. If you do remove this IC, you'll need an exact replacement. The number of the IC should match exactly, including any code numbers that might follow the part number on the chip. Otherwise, some functions may not work.

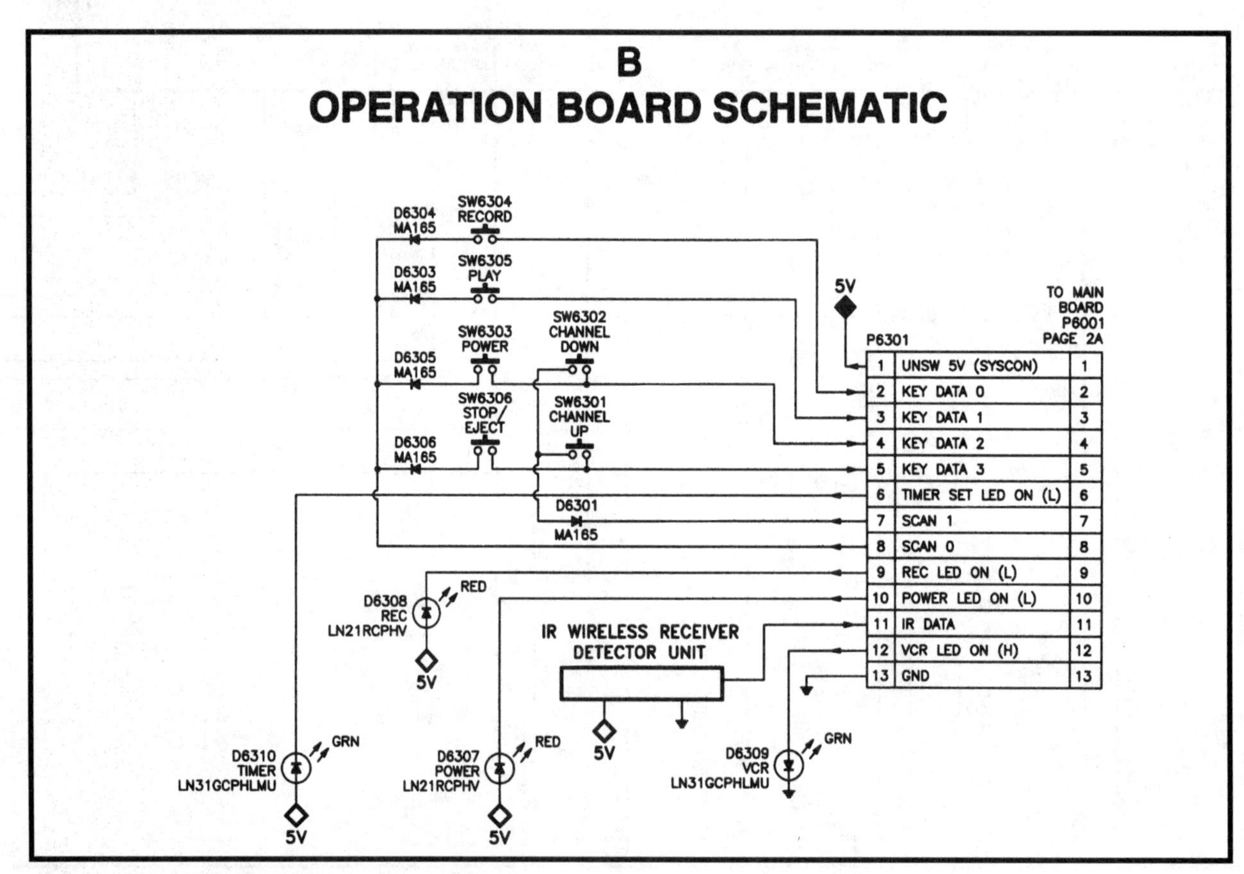

Figure 11-4. The keypad schematic of the VCR.

Chapter 12

Troubleshooting Display Circuits

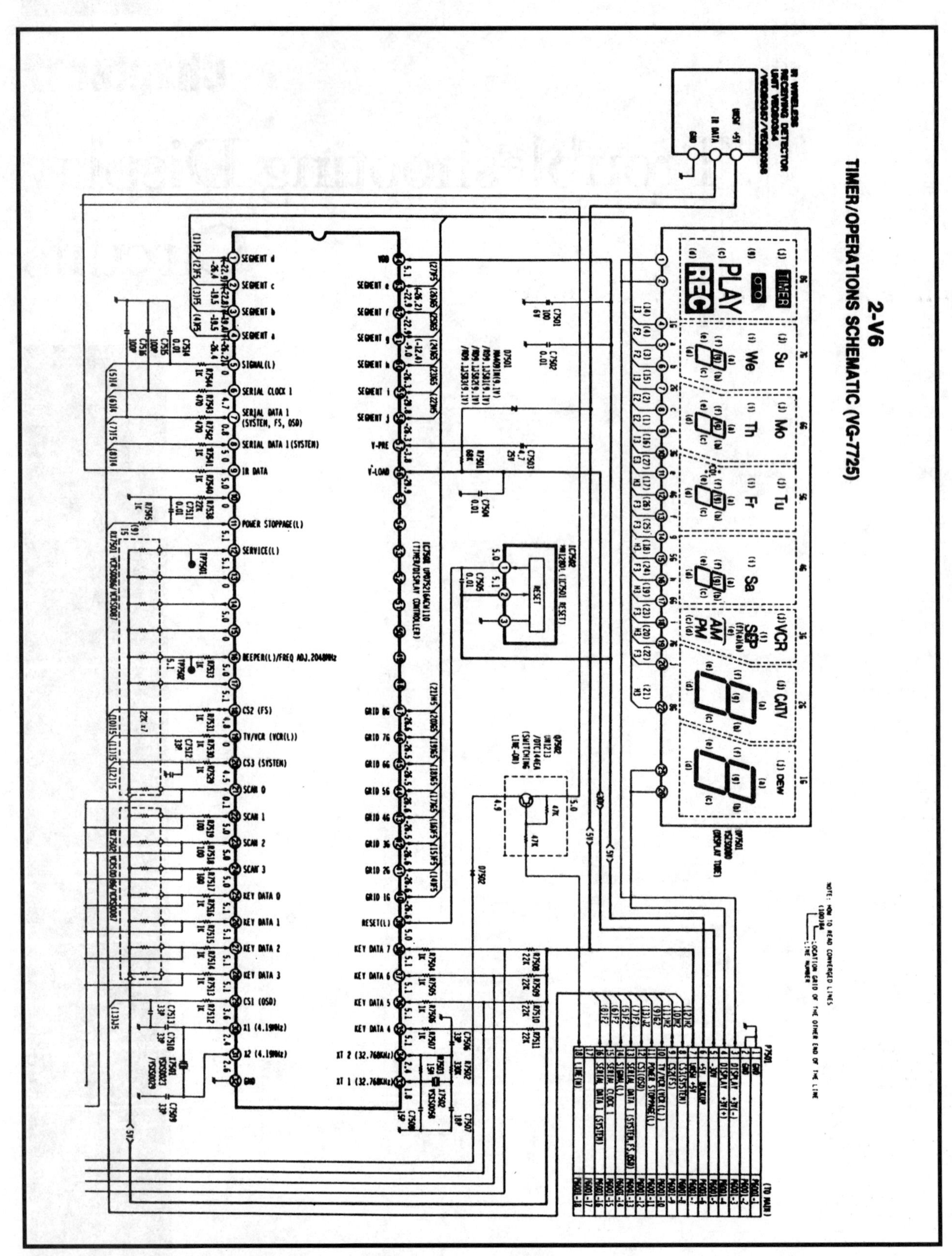

Figure 12-1. Schematic diagram of the display circuitry of the VCR. (Courtesy of Thomson Consumer Electronics)

A VCRs display circuits consist of the actual display that you see on the front of your VCR and the circuits that drive the display. The latter is usually a timer chip and its support circuitry or, on some models, the microprocessor. **Figure 12.1** shows the schematic diagram of the display circuitry of the GE VG-7725.

As mentioned earlier, the display is the first item you check when assessing the condition of a VCR. If the display is dead, it will not blink. This usually points to a problem with the power supply. If the display is blinking, the power supply is probably okay.

12.1 Checking the Display

The VCR display tells you time, channel, mode, and other information. **Figure 12.2** shows the display of the GE VCR. The display is most often damaged when a VCR is dropped or kicked. If a display is cracked, it must be replaced.

Most modern VCRs use fluorescent displays. These have filament pins on each end of the display module. These are typically powered by a low AC voltage of about 3 V. If the solder joints of these pins crack, the display will not operate. You have to closely examine the printed circuit board where the filaments are attached. If you see a cracked solder joint, resoldering the joint will solve the problem.

The entire display usually consists of six 7-segment displays for showing the time and channel as well as various icons and labels, such as days of the week. If a segment or label fails to light, you need to check the voltage on the pins that serve that part of the display. If the voltage is missing, it may be due to a bad solder joint or a problem with the driver chip.

A common problem with VCR displays is intermittent behavior of a label or icon. For example, the tape icon turns on and off in a random fashion. This problem can be

Figure 12-2. The VCR's display.

caused by a bad connection, a faulty driver chip, or a faulty sensor. Often, when the icon goes out, certain VCR functions will not work. For example, if you place a tape in the VCR the tape icon should light and stay lit until you remove the tape. If it goes out while the tape is still in the VCR, functions like PLAY will not work. This points to a problem with the sensor that signals the microprocessor that the tape is in the machine. If the sensor is operating intermittently, maybe due to a loose wire, the signal it sends to the microprocessor will go on and off rather than staying on. Every time the sensor turns off, the microprocessor will signal the display driver to turn off the icon and will not respond to the PLAY command.

Troubleshooting intermittent problems of this type demand a good knowledge of the circuitry and a lot of patience. You probably will need to refer to the service manual if you don't have experience with the particular VCR chassis.

12.2 Troubleshooting the Timer Circuit

The display driver is typically the timer chip of a VCR. **Figure 12.3** shows the timer chip of the GE VCR. Notice in this model VCR, that the display sits right on top of the timer chip. You cannot examine the chip unless you release the plastic latches of the display and rotate it up on its hinge.

Figure 12-3. The timer IC on this VCR is located underneath the display, so you have to lift the display up on its hinge to get to the timer IC.

Timer chips are large ICs, usually with more than 60 pins. The timer interfaces with the microprocessor and drives the segments of the display. If you suspect a bad timer, make sure to check whether it is receiving power. The timer chip of the GE VCR receives power at pin 64, VDD. The voltage is 5.1 V. You also need to check if the crystal is working. The timer chip in the GE VCR uses two crystals, one between pins 33 and 34 (32.768 kHz), and another between pins 30 and 31 (4.19 MHz). You can check the frequency with a frequency counter. **Figure 12.4** shows how to check the crystal with the frequency counter of a Sencore SC61 Waveform Analyzer. The scope display shows the waveform, while the LCD shows the frequency. These checks can also be made with a regular scope or a frequency counter.

If the chip is getting power and the crystals are working, the chip should be operating properly. An intermittent problem may be due to a faulty chip, but may also be caused by a faulty sensor. Replacing a timer chip requires a big soldering job and replacement with an expensive part. If you are absolutely sure that the timer is bad, you need to desolder all the pins with a desoldering tool and probably a solder wick. **Figure 12.5** shows a timer chip being desoldered. After all the solder is removed, you can pry the chip off the printed circuit board with a small flathead screwdriver. **Figure 12.6** shows how this is done.

Keep in mind that the timer chip is a sturdy circuit. It rarely goes bad. Make sure to check all other possibilities before you decide to pull this IC.

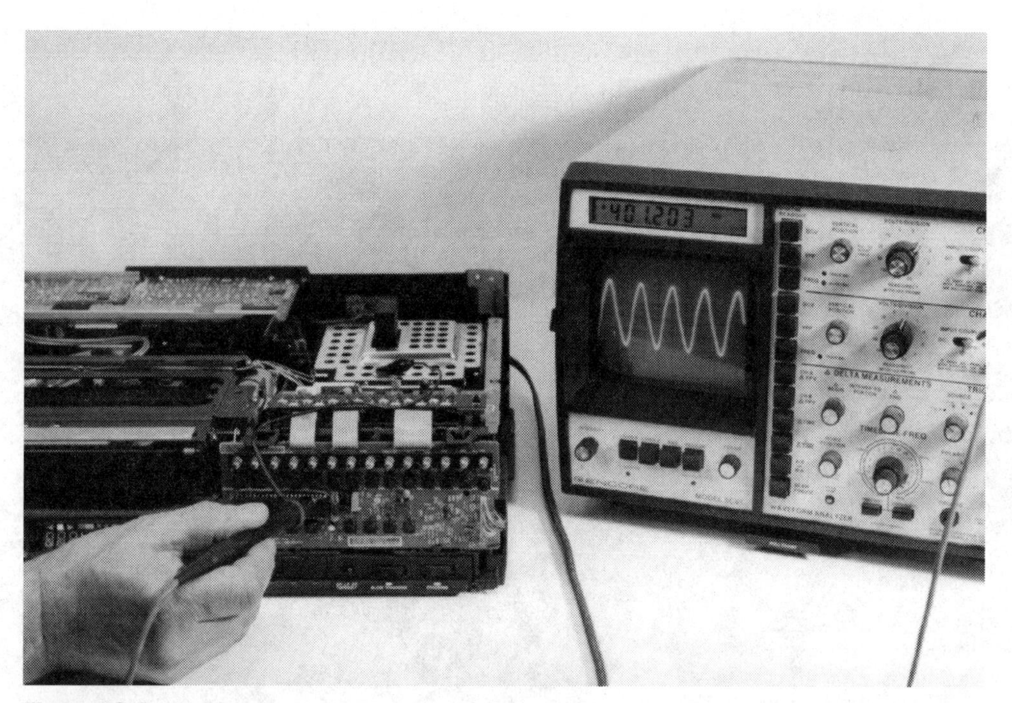

Figure 12-4. Checking the crystal with the frequency counter of a Sencore VCR Universal Video Analyzer.

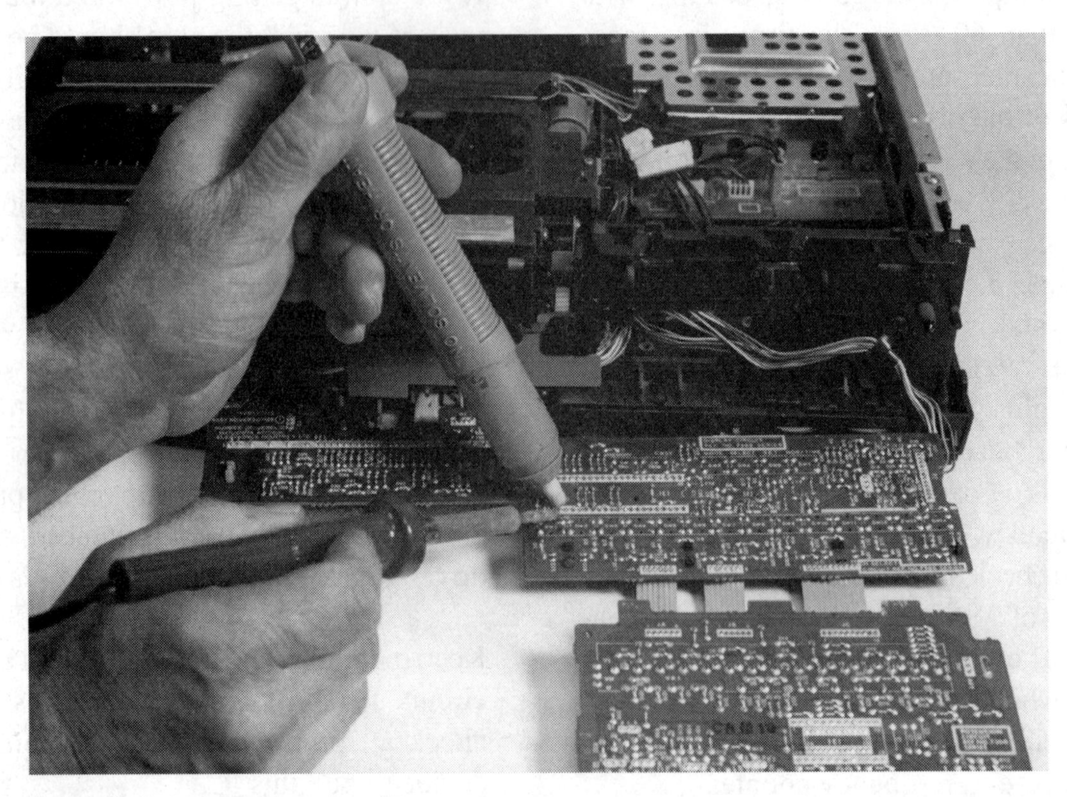

Figure 12-5. Desoldering a timer IC.

Figure 12-6. Prying the timer IC off the circuit board after desoldering.

Chapter 13
Remote Control Circuits

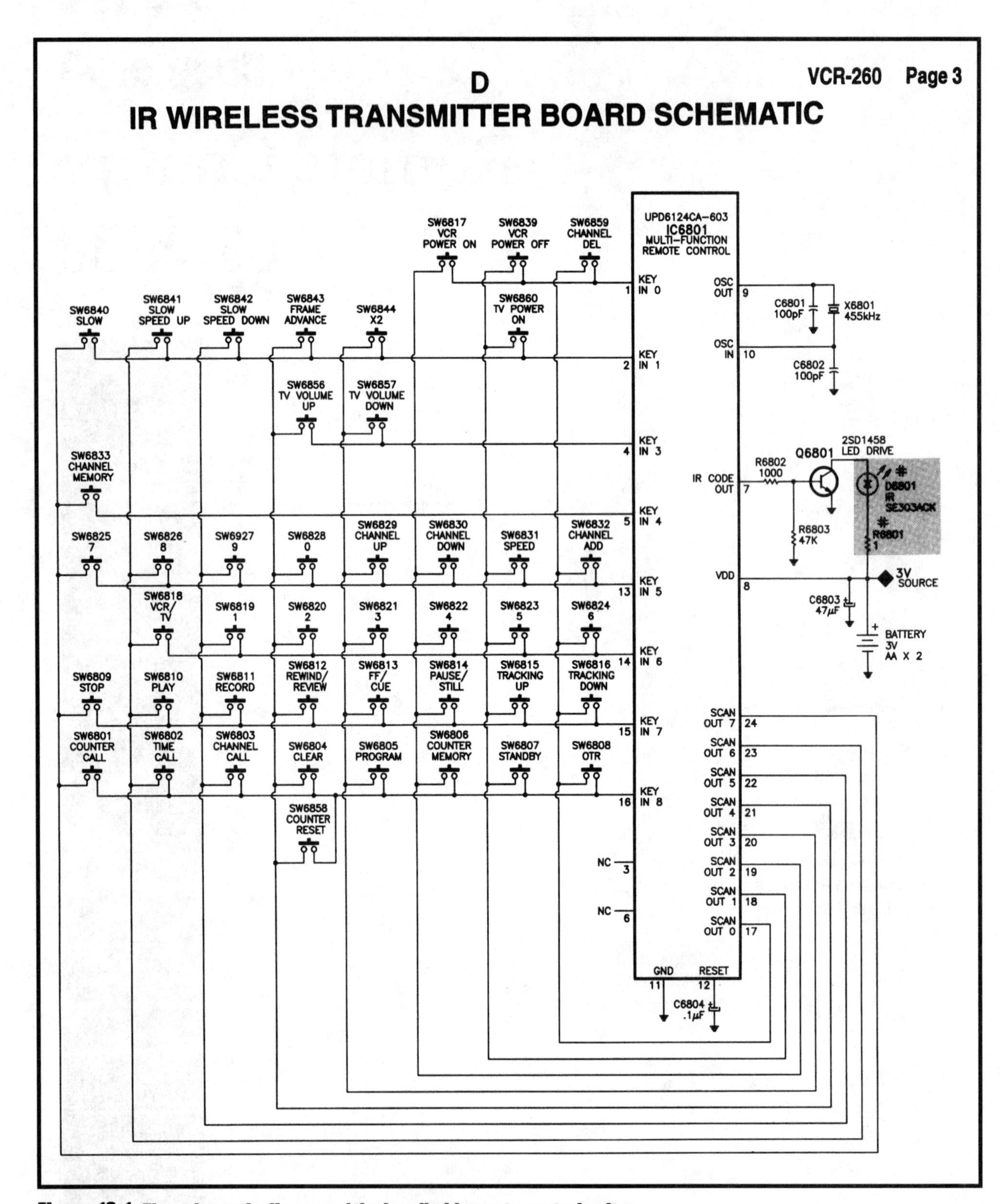

Figure 13-1. The schematic diagram of the handheld remote control unit.

The remote control circuitry is in two parts. One part is the handheld remote control unit. The other part is the receiver, which is inside the VCR, just behind the front panel. It receives the signals from the handheld unit and sends them to the microprocessor for execution. **Figure 13.1** shows the schematic diagram of the handheld unit of the GEVG-7725. **Figure 13.2** shows the interior of the handheld unit. **Figure 13.3** shows the infrared receiver.

If there is any problem with the remote control, most of the time the trouble is with the handheld unit. If there is a problem with the internal receiver, it's either a bad infrared diode or a bad connection.

Figure 13-2. The interior of the handheld unit showing the foil side of the printed circuit board on the left.

13.1 Troubleshooting the Handheld Remote Control Unit

If a remote control handheld unit dies, it's often desirable to repair it rather than to replace it with an inexpensive "universal" remote control. Obviously, the universal remote doesn't have as many functions as the original model. The original remote has programming functions and can call up the on-screen-display (OSD). Universal remotes provide only the basic functions like power, play, rewind, fast forward, fast rewind and channel selection.

Problems with the handheld remote controls are often caused by liquid spills, such as coffee and cola, or traumatic occurrences, such as being stepped on. Once a liquid gets between the rubber buttons and the printed circuit board, it stays there. The rubber cover makes it hard for the liquid to dry, and even when it does, it leaves a residue of sugar on top of the PC board. When you press the button, no contact is made.

Figure 13-3. The infrared receiver is housed in a small metal container.

The easiest way to fix a problem caused by a liquid spill is to very carefully open up the remote control unit, take the circuit board out, and wash it with water. If you use any sprays or alcohol, you may not dissolve the sugar. Water will do it. After you clean it well and make sure all the sugar is gone, then you have to dry the board completely or else it will not work. You have to do the same thing with the rubber button pad. Again, you must make sure this piece is completely dry before reinstalling it.

Another problem that crops up with a handheld unit concerns the crystal. You should examine this very carefully. When a remote control is dropped or thrown on the floor, one of the leads of the crystal may break, or the solder joint may crack. Before you resolder the crystal, you should place one drop of hot-melt glue on the body of the crystal to secure it to the PC board. This will help it to better withstand shocks in the future.

When you examine the PC board, watch carefully for any broken traces. Sometimes, the remote gets stepped on. This can break a trace or the board itself. If the board is shattered, obviously, it doesn't pay to try to fix it. Then you are forced to use a universal remote or order a replacement from the manufacturer.

If a close inspection of the handheld unit shows no abuse, the problem may be with the infrared diode that sends the pulses to the VCR. If this is defective, it has to be replaced with an original component. Also, the transistor that amplifies the signal from the remote control IC and sends it to the diode should be checked out. If this transistor is open or shorted, the remote control will not work.

13.2 Troubleshooting the Remote Control Receiver

As far as the receiver goes, there is a simple test to check whether or not it is receiving pulses from the handheld unit. You can perform this check with a DMM without taking the receiver apart. The receiver unit has only three leads, power (5 V), ground, and the output. Set the DMM to the 10 to 12 VDC range and connect it to the output of the receiver. Then, hold the handheld unit against the face of the receiver and push any button. You should see a reaction at the output. If you want to watch the pulses go by on a scope, you need to press a continuous function button such as the volume, rather than a momentary function like PLAY. If you are getting a signal at the output of the receiver, chances are both the handheld unit and the receiver are working properly.

The handheld unit and the receiver may be working properly (as checked with a DMM) and still the VCR may ignore the commands. How does this happen? The problem may lie in the connection between the receiver board and the main PC board of the VCR. The connection may be broken or dirty or have some other problem that affects continuity. If this happens, repairing the break or cleaning the plug will solve the problem.

Chapter 14

Troubleshooting Servo Systems

There are two servo systems in a VCR, the cylinder servo and the capstan servo. The purpose of the cylinder servo system is to control the speed and phase of the video heads. Cylinder speed control assures that the cylinder or drum motor spins at exactly 1,800 revolutions per minute or 30 Hz (revolutions per second) when recording. Cylinder phase control assures that the video heads are correctly positioned over the appropriate track on the video tape, with relation to motor speed.

The purpose of the capstan servo system is to control the speed and phase of the video tape. Capstan speed control assures that the tape moves at the correct speed during playback. Capstan phase control assures that the positioning of the tape is correct, in relation to the video heads.

14.1 Overview of the Servo Systems

In the GE VG-7725 VCR these two servo systems are implemented with five ICs as shown in the schematic diagram of **Figure 14.1**. **Figure 14.2** shows the servo control section on the main PC board of the VCR. All of these ICs are relatively easy to get to. If any are found defective they need to be removed.

IC2001 (**Figure 14.3**) is the capstan/cylinder servo processor. Cylinder phase and speed control signals are at pins 19 and 20, respectively. Capstan speed control and phase comparator are at pins 21 and 22, respectively. Among other things, this IC is responsible for tracking control in the VCR.

IC2003 and IC2005 (**Figure 14.4**) are the motor drive ICs for the cylinder and capstan motors, respectively. IC2004 (**Figure 14.5**) is the capstan main coil drive IC. IC2002 is the capstan servo slow/still processor. This IC is used to implement the slow motion and pause functions of the VCR.

A variety of signals are used for servo control in a VCR. In the GE model that we are covering, the capstan FG (frequency gen-

Figure 14-2. The servo control section on the main PC board.

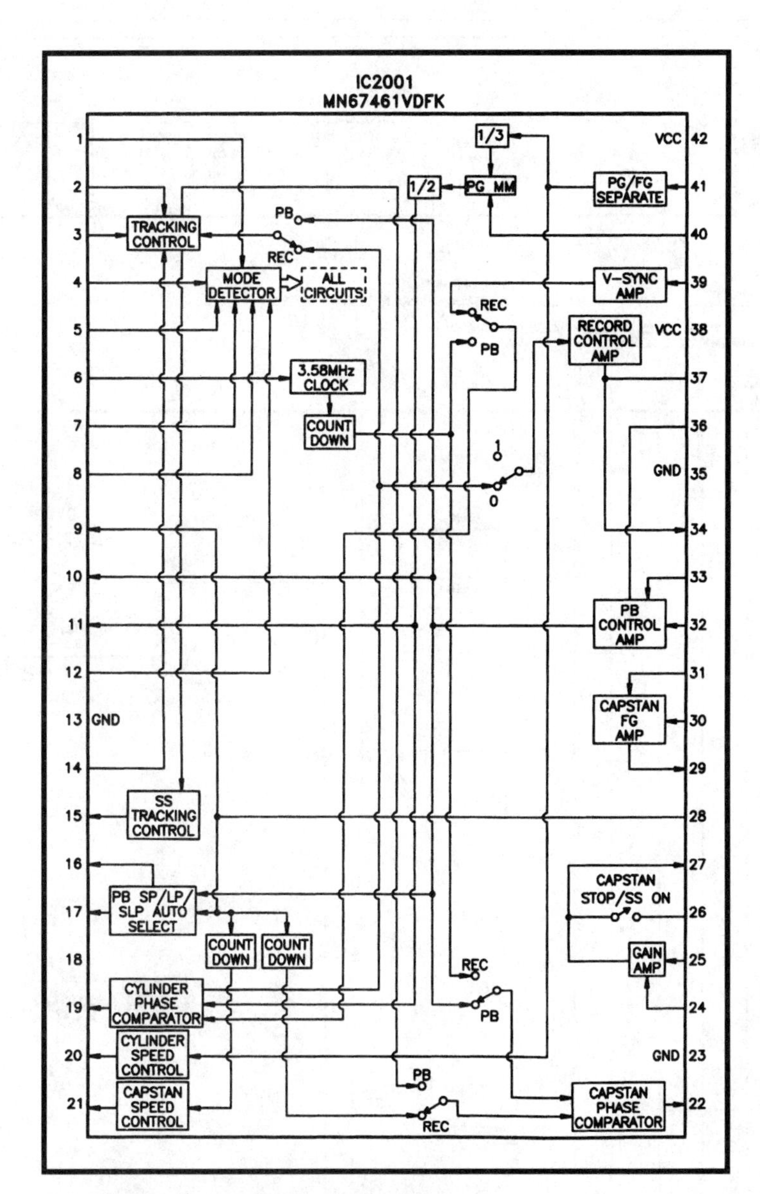

Figure 14-3. Block diagram of IC2001, the Capstan/Cylinder Servo Processor.

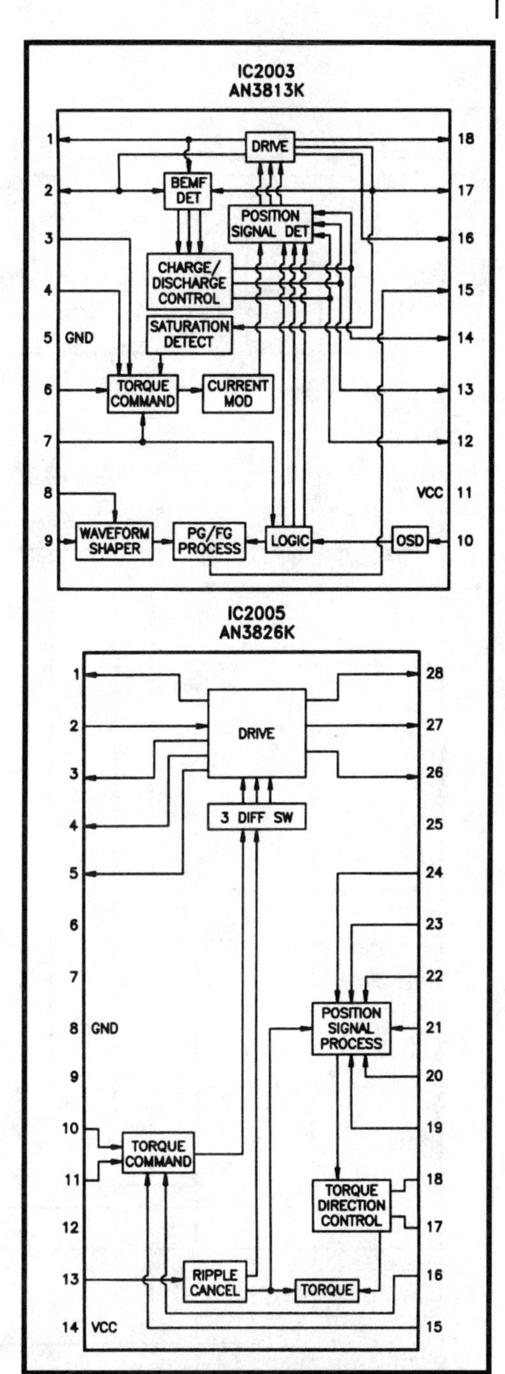

Figure 14-4. Block diagrams of IC2003 and IC2005, the Motor Drive ICs for the cylinder and capstan motors, respectively.

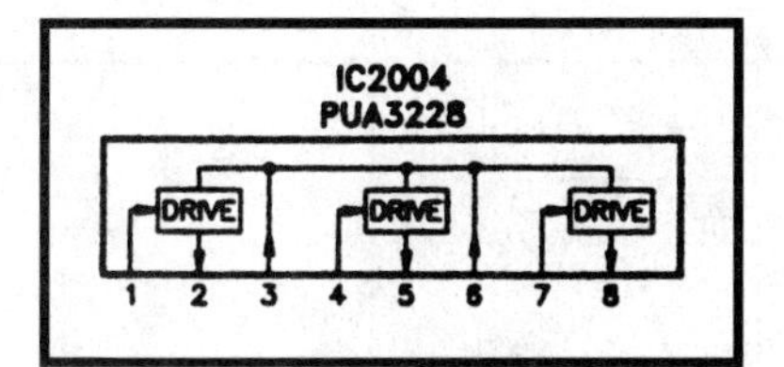

Figure 14-5. Block diagram of IC2004, the Capstan Main Coil Drive IC.

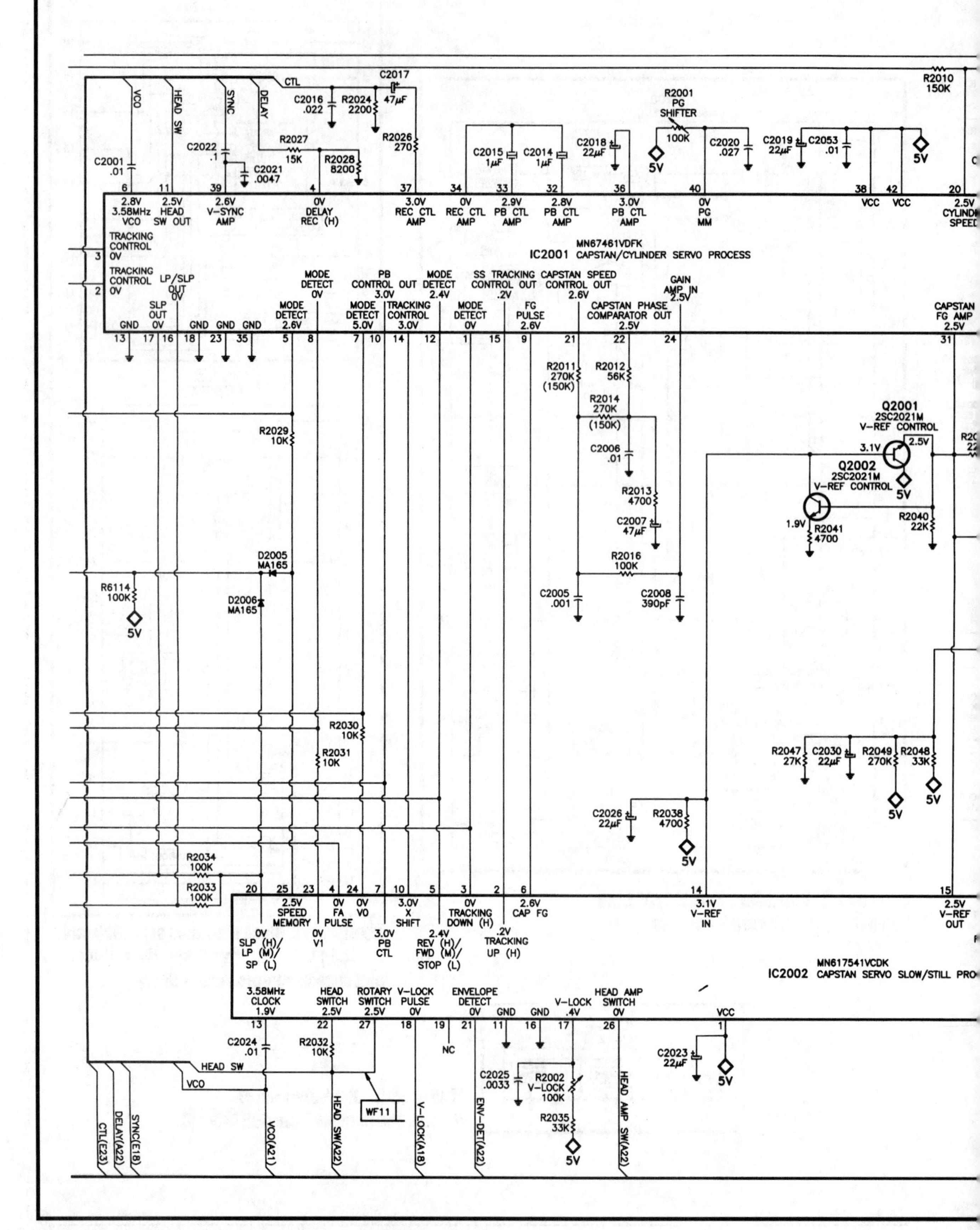

Figure 14-1. Schematic diagram of the servo systems of the VCR.

D

HEMATIC continued

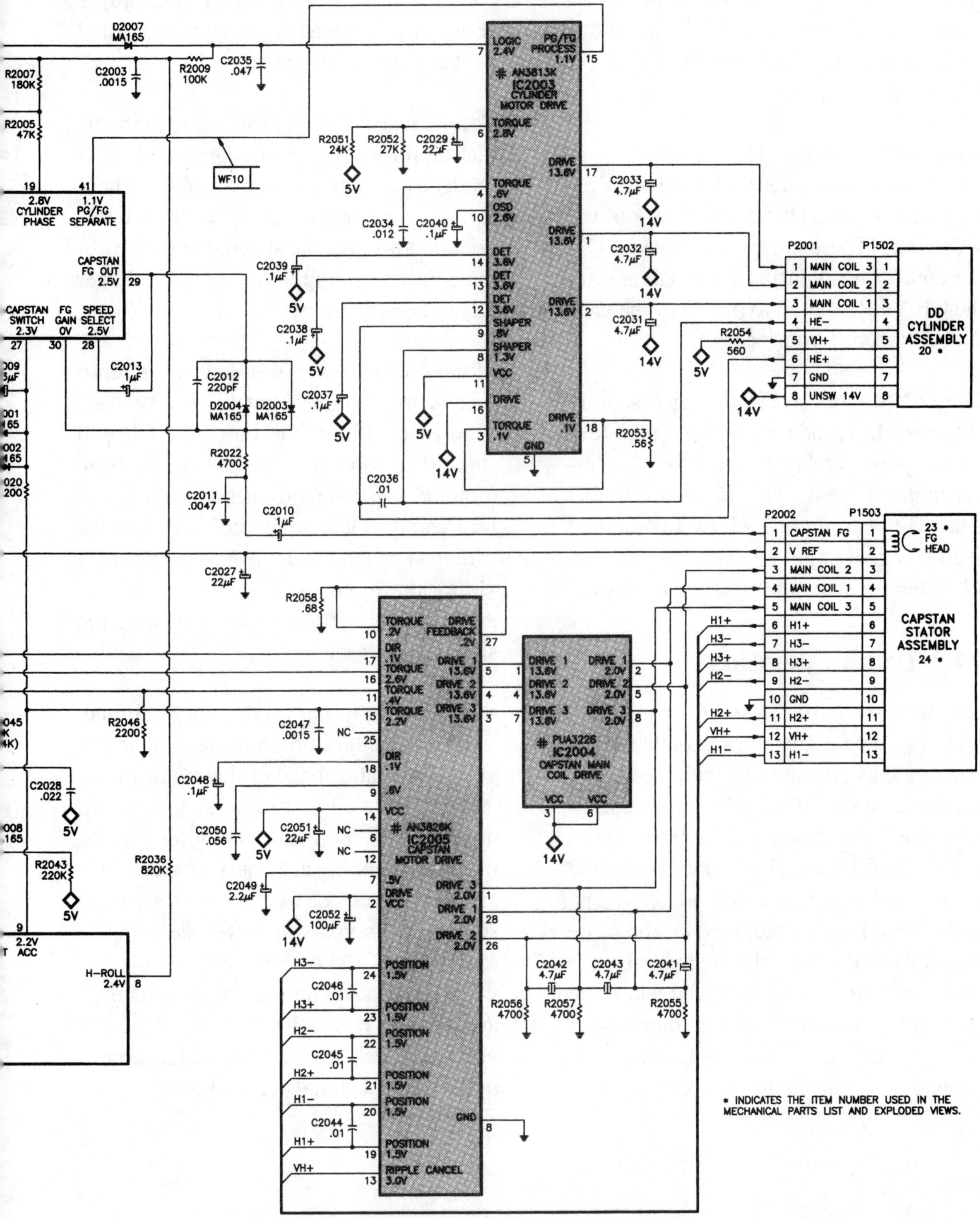

P2001		P1502
1	MAIN COIL 3	1
2	MAIN COIL 2	2
3	MAIN COIL 1	3
4	HE−	4
5	VH+	5
6	HE+	6
7	GND	7
8	UNSW 14V	8

DD CYLINDER ASSEMBLY 20 *

P2002		P1503
1	CAPSTAN FG	1
2	V REF	2
3	MAIN COIL 2	3
4	MAIN COIL 1	4
5	MAIN COIL 3	5
6	H1+	6
7	H3−	7
8	H3+	8
9	H2−	9
10	GND	10
11	H2+	11
12	VH+	12
13	H1−	13

23 * FG HEAD

CAPSTAN STATOR ASSEMBLY 24 *

IC2003 AN3813K CYLINDER MOTOR DRIVE

IC2004 PUA3226 CAPSTAN MAIN COIL DRIVE

IC2005 AN3826K CAPSTAN MOTOR DRIVE

* INDICATES THE ITEM NUMBER USED IN THE MECHANICAL PARTS LIST AND EXPLODED VIEWS.

erator) pulses are developed in the capstan stator assembly (**Figure 14.1**). In 2-hour mode, the FG signal is 720 Hz, in 4-hour mode, it is 360 Hz, and in 6-hour mode, it is 240 Hz.

The 30 Hz control (CTL) track pulses are the vertical sync pulses of the broadcast signal recorded on the control track of the video tape. These pulses are picked up by the control head during playback and, in the GE VCR, are fed to pin 37 of the capstan/cylinder servo processor.

The DD (direct drive) cylinder assembly (**Figure 14.1**) produces the PG (pulse generator) pulses and feeds them to the cylinder motor drive IC. This IC mixes the PG and FG pulses and develops a combined PG/FG signal at pin 15. This is fed to pin 41 of the capstan/motor servo processor.

14.2 Troubleshooting

For the most part, the servo system in a VCR is very reliable. One possible problem you may encounter is with the speeds available on the VCR. A VCR sometimes has trouble determining the correct speed to play (SP/LP/EP). This happens when the audio head is not mounted properly. The audio head is often mounted on the same assembly as the control head, which reads the information on the control track. If the audio head is misaligned, the control head will be, too. The VCR cannot read the information on the control track of the tape and thus has a problem determining the correct speed, which is recorded on the control track. Head alignment is essential for keeping proper speed.

You may also have a problem with the IC that amplifies the servo signals. In older models, op amps are used to amplify these signals. If an op amp goes bad, the VCR is not able to select the speed. On the newer machines, this amplifier is part of a larger IC, and rarely goes bad.

Motors, which are controlled by the servo system, may burn out, but this is not a usual occurrence. If a motor dies, the VCR will shut off as soon as it senses that the drum motor or capstan motor is not turning. Before you decide that a motor is bad, make sure to check that the power supply is providing the motor with sufficient voltage and that the microprocessor is sending the signal to start the motor.

Sometimes a motor drive IC will go bad. Evidence of this problem, as you might guess, is erratic speed of the motor that is under control. For example, you might notice the video drum changing speeds, instead of running at a constant speed. It is always a good idea to make as many measurements as you can before deciding to remove an IC. Make sure to check that the voltages (as shown in the service manual) are correct. If you have an oscilloscope, you can check whether the IC has appropriate control signals at each pin.

Chapter 15

Troubleshooting Video Heads

Determining if the video head is bad without any special equipment is very difficult. You need equipment such as the Sencore analyzer described earlier in the book. If the video head is not producing a signal, this equipment allows you to inject a video signal into the top of the video head. If you see a test pattern on the monitor, you can be certain that all the circuitry behind the video head is working. Then, you know that the video head is to blame for the problem.

With this said, we will now tell you the best ways to determine the condition of a video head—without using expensive equipment. If you are lucky, you will discover some flaw in the video head that tells you for certain if it needs to be replaced.

15.1 Checking Video Heads

A video head is a very narrow, sharp edged head that is very fragile and easy to break. If you use a cotton swab to clean the head, for example, chances are you will break the head. A video head is an integral part of the upper video drum assembly (**Figure 15.1**). Most VCRs have either two or four heads, but some have three or five. Even though the drum and the heads are all part of one assembly, most people refer to the entire unit as the video head.

Figure 15-1. A video head is an integral part of the upper video drum assembly; usually there are two or four heads.

You can better understand the condition of a video head just by looking at it under a magnifying lamp or glass. The video head should look sharp and it has to have a smooth surface. If any part of the surface is chipped, scratched or broken, the head is no good. Then, you have to replace it with a new assembly to make the machine operational.

You can check for a worn video head by applying pressure on the left side of the tape guide. Putting your finger there increases the tension on the tape. If you obtain a picture on the monitor, it means that the head is worn out. When the head is worn out, it doesn't reach the tape to read the signal on the tape. If this is the case, the video head has to be replaced.

You will sometimes find a small scratch on the surface of the drum. This scratch cannot be removed, cannot be polished. The scratch will scrape oxide from the tape and create a cushion so that eventually the head will not read the tape. If there are any scratches on the video drum, the assembly has to be replaced.

After a VCR has been used for many years, the video head may wear down but still produce a good picture. But another problem can occur. The surface of the drum has a series of tiny lines or grooves all around it (**Figure 15.2**). These grooves serve a purpose. They prevent the video tape from adhering to the video head drum,

Figure 15-2. The grooves on the video drum prevent the video tape from adhering to its highly polished surface.

which is highly polished aluminum. The grooves provide a little cushion of air for the head when it is spinning at 1,800 rpm. This little air cushion reduces the tension on the tape. The video tape slides smoothly without any jitter. If the video heads are still good but the grooves are worn out, the VCR heads will need cleaning every time the machine is used. After a couple of hours of use, the picture suddenly gets lost. You clean the heads, and you get the picture back. In another few hours, the heads get dirty again. If this happens, the video head assembly needs to be replaced.

15.2 Checking Head Switching Circuits

Head switching circuitry is a very reliable part of a VCR. This circuitry switches from one head to another, as the heads are spinning, making sure that the head in contact with the tape is connected to the amplifier. Head switching circuitry is contained in the head amplifier section of the VCR. This section is encased in a metal shield and can be found right behind the video heads (**Figure 15.3**). The switching frequency is taken from the frequency of a 3.58 MHz crystal in the VCR.

Some very old machines use a relay that switches both pairs of heads. Modern VCRs use a solid state switch that switches the heads depending on the speed of the drum motor. The GE VG-7725 uses the AN3334K IC shown in **Figure 15.4**. As shown in the schematic diagram of **Figure 15.5**, the HEAD SW signal is at pin 1 of this IC. You very rarely find any problems with this switch.

Figure 15-3. Head switching circuitry is contained in the head amplifier section of the VCR, usually found right behind the video heads.

15.3 Replacing Video Heads

If a video head is found to be defective, you have to change it. This is comparatively easy. There are only two ways to place a video head in a VCR, the correct way and 180 degrees out of alignment. If you have a 5-head machine, you need to be more careful to see that the new head aligns in the machine the same way as the old head.

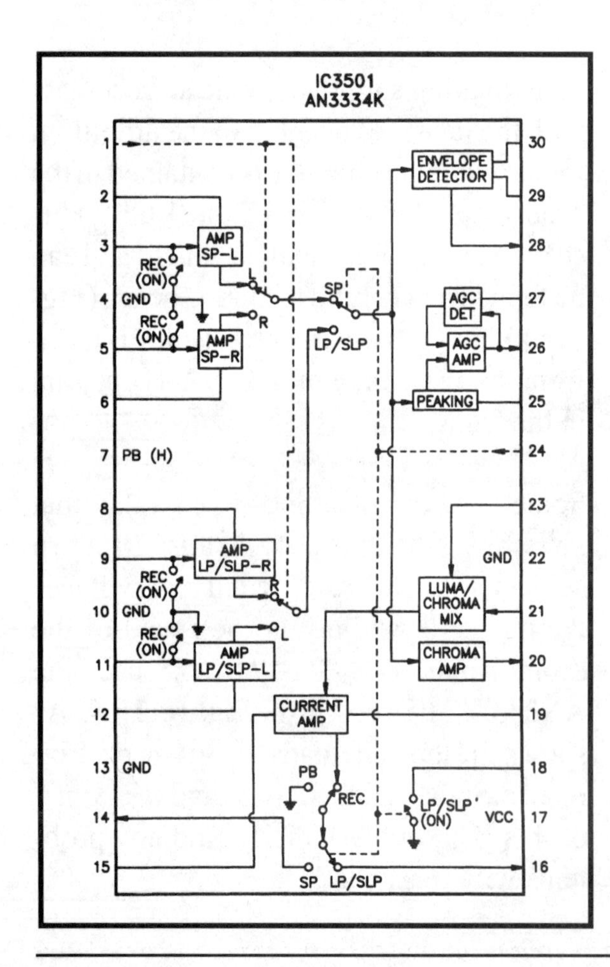

Figure 15-4. Block diagram of the AN334K IC, used for video head switching.

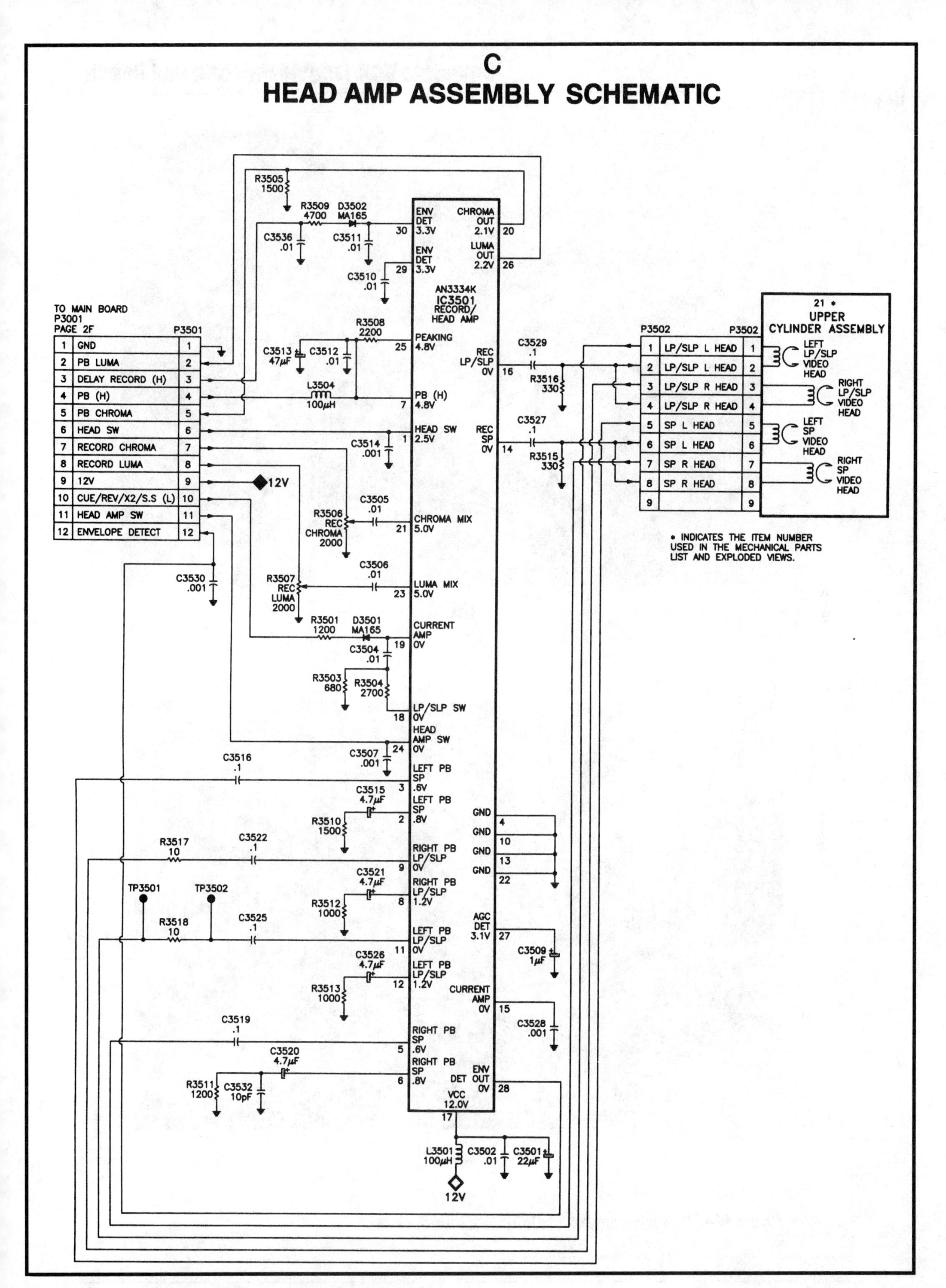

Figure 15-5. The HEAD SW signal is at pin 1 of the AN3334K IC.

[A]

[B]

Figure 15-6. Removing the anti-static bracket from the VCR.

The video head is attached to the shaft with only two screws. Marking the shaft and the old drum will help you put in the new part correctly. The disassembly procedure starts by removing an L-shaped metal bracket over the video drum. This bracket prevents the buildup of static electricity. You have to remove the screw that attaches this bracket to the chassis (**Figure 15.6a**) and then remove the bracket (**Figure 15.6b**). Next, you must remove the two screws that secure the upper video drum to the rest of the assembly (**Figure 15.7**). Then, you have to desolder the wires that go to the head (**Figure 15.8**). On some VCRs, this step is not necessary. Connectors from the transformer on the bottom of the video head have pins that go to the video head printed circuit board. These heads fit into the machine only one way.

After you desolder the wires from the heads, you have to pay close attention to the colors of the wires. You don't want to reverse the wires when you resolder them to the new head assembly. At this point, you can remove the old head (**Figure 15.9**). Before you seat the new head, make sure that it is clean, with no lint, dust or debris. Put the head assembly in carefully, gently pressing it down. Don't force it or bend it. When it is seated correctly, replace the two screws and make sure you tighten them.

No further alignment is necessary on modern VCRs. Once the head is put in properly and the screws tightened, there is no need for further alignment. You may now align the tape guides, if necessary, but you don't have to align the video heads.

Figure 15-7. Removing the two screws that secure the upper video drum to the rest of the assembly.

Figure 15-8. On some VCRs, you have to desolder the wires that go to the video heads.

Figure 15-9. Removing the upper video drum.

Once you install the new heads, you should check the PG shifter, to make sure you are not getting lines at the bottom of the screen. Then, you should check the alignment of the left and right tape guides. You have to be able to play a tape recorded in another machine with only a minor adjustment of the tracking control. If a VCR has automatic tracking, it has to be able to play any tape recorded on any machine (unless the machine that recorded the tape is out of alignment).

15.4 Aligning the Tape Guides

Aligning the tape guides is quite simple. To do this you need a test tape and an oscilloscope. You can purchase a test tape from the manufacturer, but these are very expensive. To avoid this expense, you can make a test tape yourself. If you have access to a new 4-head VCR, make a 5-minute recording of color bars or a 10-step signal. It is fairly easy to make a test tape if you have a VCR in good condition.

Once you make the test tape, put it in the VCR that you are aligning. If you get an acceptable picture, that is, the picture is free of noise and is clear, then all you have to do is check the range of the tracking, from left to right. On the left side of the tracking control, you have to lose the picture. On a properly aligned VCR, the picture will be snowy all over, not only on the top and bottom or only on the top. The picture must deteriorate gradually, not in one part, but all over the screen. If this happens, then the VCR has good alignment. If it doesn't happen, the VCR needs to be aligned. Also, if you have to turn the tracking all the way to the left or the right to obtain an acceptable picture, the VCR needs to be aligned.

Put the test tape into the VCR. Then, connect an oscilloscope to the VCR. You can't do the alignment without a scope. Even if the scope is not very sophisticated, it can do the job. You connect the scope to the FM test point in the VCR and press PLAY. You need to synchronize the scope until the pattern on the screen stops moving, or you can use an external trigger. You can trigger the scope from the 30 Hz head switching signal. Then, the waveform will lock. When you do the alignment, if there is a gap to the right side of the waveform, that means the outgoing tape guide, which is on the right side, has to be aligned. You move it either up or down until you get a straight line on the scope. If the left side of the waveform drops out, then you have to align the left tape guide.

Sometimes, when you do an alignment, you can't get the line straight no matter what you do. The problem is that the tape posts deteriorate over the years. Or sometimes they have oxide deposits from the tape, which changes the height of the tape post. If you have this problem, you have to clean the tape post very carefully. Also, you should clean the pinch roller and the audio head very carefully. The position of the audio head has a lot to do with the alignment of the video head. If the alignment is done properly, but you still cannot bring the tracking to the center, then you have to move the audio head slightly to the left or right.

Chapter 16

Troubleshooting Power Supply Circuits

Of all the circuitry in the VCR, the power supply is by far the source of most electrical problems. Why is this so? The power supply is subject to many of the conditions that make electronic components unreliable. Electrical surges and heat are two prime examples. The power supply usually takes the brunt of any electrical surge in the AC line. It takes the first punch, in a sense. Heat buildup in the power supply often dries out capacitors, causing them to fail. If you turn on the power of a VCR and see no response (the display does not come on), check whether the machine is plugged in. If it is, you very likely have a problem with the power supply.

16.1 Linear Power Supplies

Linear power supplies are an older type of supply, but are still used in newer VCRs. **Figure 16.1** shows the schematic diagram of the linear power supply in a Sharp VC-2210U VCR. Linear power supplies are not very efficient and weigh a lot, but they are easy to build and reliable.

The linear supply employs a standard step-down transformer. The transformer serves two functions. It lowers the voltage, and it provides isolation between the primary side, which is connected to the AC line, and the secondary side. If more than one voltage is needed by the equipment, the secondary of the transformer will have several secondary windings that produce different voltages. This is the case with all VCRs.

On the secondary winding side of the transformer can be bridge, full-wave and half-wave rectifiers. The bridge rectifier is the most efficient, allowing a smaller filter capacitor to be used. Most VCRs with linear supplies use all three kinds of rectifiers.

A bridge rectifier is constructed with four diodes. These diodes may be discrete, as in the Sharp supply (D902-D904), or housed in a single package that looks like an integrated circuit. A full-wave rectifier is constructed with two diodes, D908 and D909 in the Sharp supply. A half-wave rectifier employs just one diode, D911 in the Sharp supply. All three types rectify the AC and produce a DC voltage. This voltage is filtered by a capacitor and then fed to a regulator, which regulates the outgoing voltage.

The supply may have one or more regulators depending a the number of voltages required for operation of the equipment. The Sharp supply of **Figure 16.1** produces six different voltages. These voltages are applied to the various circuits in the VCR.

If you notice that the fuse is blown on the primary, it means that the transformer is taking excessive current. It is not wise to just replace the fuse and power the VCR again. It's a good idea to check the full-wave rectifier and diodes and zener diodes on the secondary side of the transformer. Sometimes the zener diodes short out and create the excessive current that blows the fuse.

Normally, after the rectifier you have a filter capacitor rated at 2,200 uF. If this capacitor fails, it can cause 60 cycle hum to

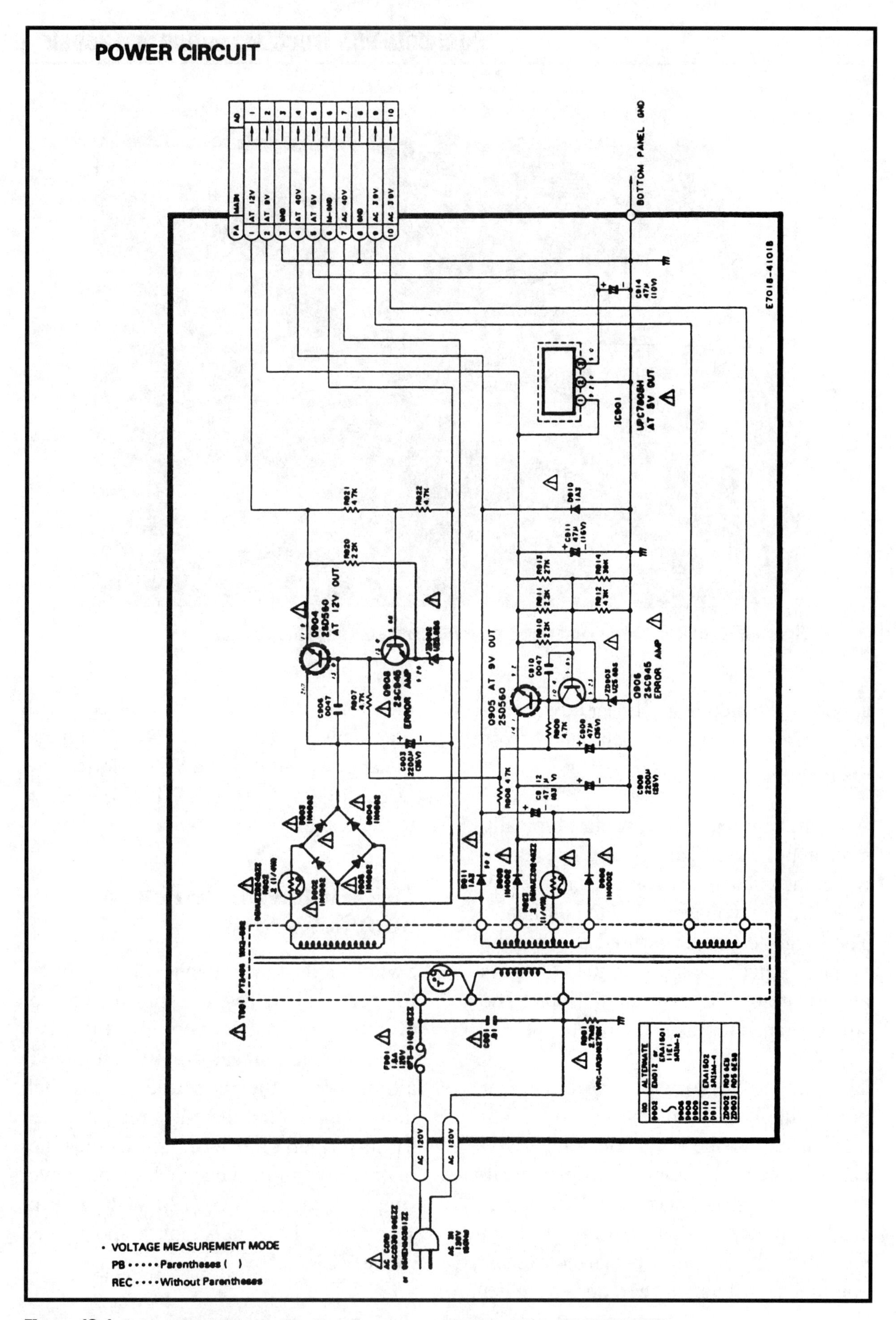

Figure 16-1. Schematic diagram of the linear power supply in the Sharp VC-2210U VCR.

Figure 16-2. A typical linear power supply in a VCR; notice the large transformer.

appear in the picture in a linear power supply. This looks like a 2-inch band across the picture.

After the regulator is a capacitor typically rated at between 10 and 50 uF. The reason is that the voltage that comes from the rectifier is a raw voltage; it has to be filtered. The bigger capacitor is used to filter the voltage. After the regulator, the voltage is almost smooth but has a slight ripple. The smaller capacitor takes the ripple out.

The things that go wrong with a linear power supply are fairly easy to spot. With a DMM set to the ohmmeter range, you can measure each component separately to see if there is any short. And the repair work is pretty straightforward. It is not something that requires special tools or test equipment. All you have to do is desolder the shorted component and replace it with a new one. **Figure 16.2** shows a typical linear power supply in a VCR. Notice how large the transformer is. This contributes to the weight of the supply.

16.2 Switch Mode Power Supplies

Switch mode power supplies also use transformers, called flyback transformers or switch mode transformers. But there is a difference. In a linear supply, first there is a transformer, then a rectifier. In a switch mode supply, first there is a rectifier, then a transformer. **Figure 16.3** shows the schematic diagram of the switch mode power supply in the GE VG-7720 VCR. **Figure 16.4** shows how the supply is positioned in the VCR.

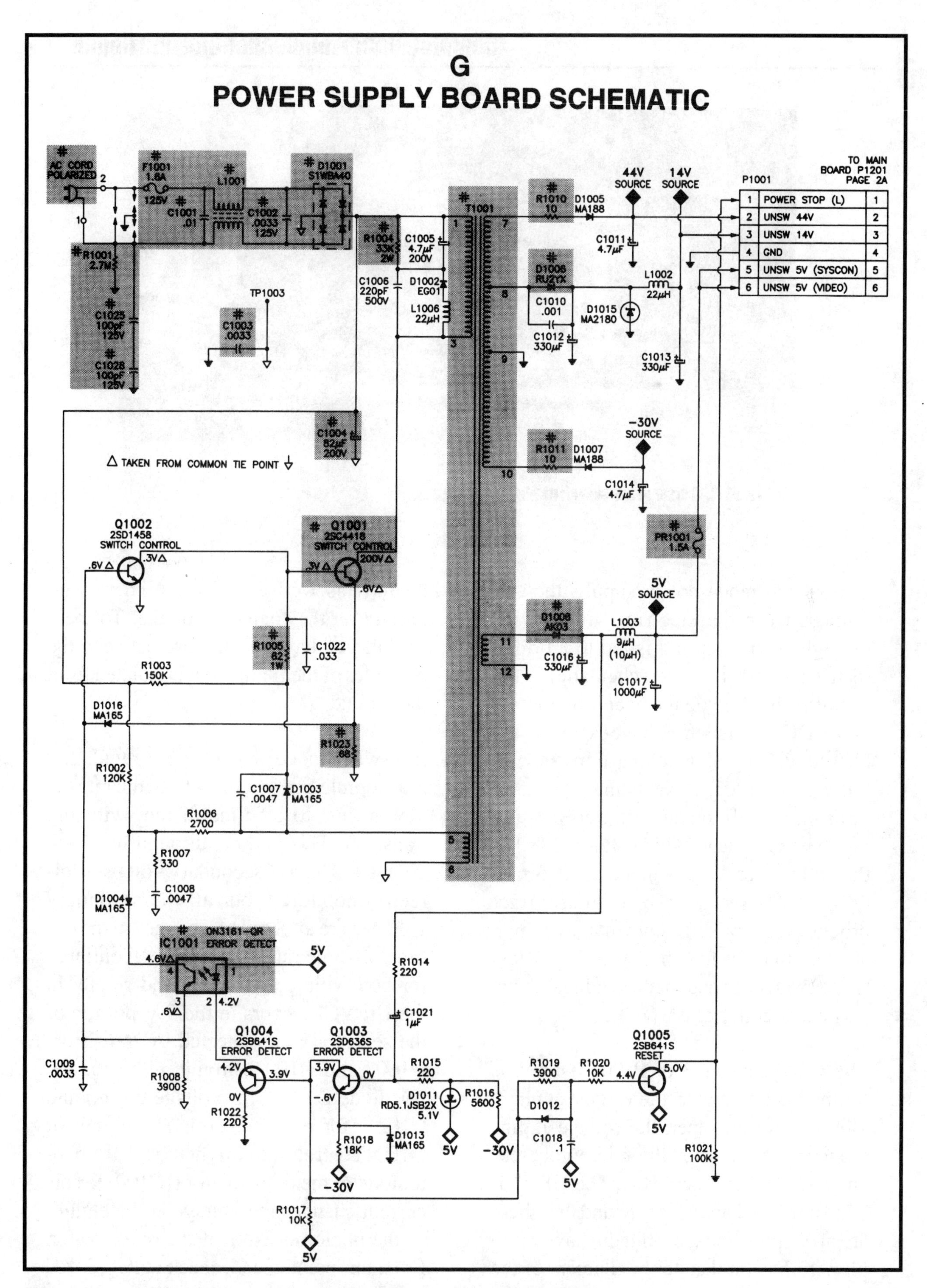

Figure 16-3. Schematic diagram of the switch mode power supply in the GE VCR.

Figure 16-4. A view of the switch mode power supply.

In a switch mode power supply, the AC voltage from the electrical line passes through an rf filter and is fed to a bridge rectifier (D1001 in the schematic). This converts the AC voltage to approximately 150 VDC. This high voltage is converted back to AC by a switching transistor, in this case, Q1001. A switching transistor turns on and off, on and off, thus producing the AC voltage. This voltage is fed to the flyback or switch mode transformer (T1001). The secondary of this transformer produces several different voltages, which are rectified by four half-wave rectifiers (D1005-D1008) and made available to the various circuitry of the VCR.

The voltage at pin 3 of P1001, +14 V, is for the loading motor; the voltage at pin 2, +44 V, is for the tuner; the voltage at pins 5 and 6, +5 V, is for the microprocessor and video ICs, respectively. The GE VG-7720 does not have a display, and thus there are no display voltages. If it did have a display, +3 V would go to the filament of the display, as well as -30 V. There's no ground on the filament voltage. To measure it, you have to measure between the two sides of the filament. It is isolated from the ground.

In a switch mode power supply, regulation is accomplished by controlling the ratio of the on-time to off-time of the switching transistor. This is done through feedback circuitry. The 5 V secondary voltage is directly monitored, but all the secondary voltages are affected by changes in the on-time/off-time ratio. The 5 V is monitored for both voltage deviation and ripple. In the GE VCR, errors in the 5 V portion of the secondary are detected by transistors Q1003 and Q1004 through reference zener diode D1011 (for voltage errors) and C1021 (for ripple errors). The output of Q1004 controls the current though the photodiode of the optoisolator (IC1001). This current changes the voltage at the emitter of the phototransistor of the optoisolator. From this point, the feedback voltage is fed to the base of the switching control transistor (Q1002), which controls the regulation of the on and off cycle of the switching transistor (Q1001). The pulse width of transistor Q1001 is controlled in this way. This technique is called pulse width modulation (PWM). Controlling the pulse width determines how long the transistor will stay on and how long it stays off. This method adjusts the voltage in the secondary to be at a constant level, which is preset by the reference voltage.

An optoisolator is used because the primary of the power supply has a hot ground, which is connected to the electrical ground. The secondary has a separate ground. There is no way to communicate between the two grounds. The only way is with an optoisolator, which has a very high isolation voltage, about 5,000 V. The feedback works through the optoisolator to adjust the outgoing voltage.

Why do you need feedback? When the switching transistor is on, energy is stored in the transformer. When the transistor is off, the energy from the transformer is released. This is the way the switch mode power supply works. When the load connected to the secondary increases somehow (another motor is activated), the switching control transistor extends the time of the pulse so that the transistor stays on longer. When the current drops down again, the on-time of the pulse is decreased, so the transistor stays off longer.

The difficulty in servicing switch mode power supplies stems from the fact that they are so-called closed-loop circuits, just like the servo circuitry, because they use a feedback voltage. A typical problem is a blown fuse. Replacing the fuse rarely solves the problem. Something caused the fuse to blow. First of all you have to check the bridge rectifier. If this is okay, then you check the switching transistor. The switching transistor can be damaged by lightning, by spikes in the electrical line, or by overloading. If the switching transistor, Q1001, checks good, then you may put in a new fuse or a light bulb (in place of the fuse) to see if the supply is going to work.

A switch mode power supply is a self-contained unit in many VCRs. You may take it out of the VCR and service it separately on the bench. You don't need the VCR, because all the voltages produced by the power supply are fairly independent from the load. If you use the light bulb and there is a problem in the feedback loop, the light bulb will light a little bit. In a good working power supply, the light bulb is very very dim, you can hardly see any light.

If there is no voltage in the secondary, then it is a little more difficult to service the supply because all the voltages are interrelated. You have to start checking component by component, there is no other way. You check the switching transistor, you check the error detector, you check the voltage detector, power stoppage transistor, all the rectifiers in the secondary. There must be a reason that the power supply won't turn on. Normally, there is a short in the secondary so the switching transistor never turns on.

Most likely, one of the secondary rectifiers has shorted out, burned up a resistor, and destroyed the switching transistor. Sometimes, but rarely, the optoisolator goes bad or one of the error detector transistors. Capacitors sometimes dry up and don't rectify the secondary voltage, so there is no feedback coming to the optoisolator. The capacitors in the secondary are most likely to fail. The capacitors are warmed by the heat from the transformer and the heatsink of the switching transistor when the VCR is operating. After years of operation, they tend to dry out. When they dry out, the voltage drops down and the reference voltage gets lost. Also, the whole power supply is often housed in a metal chassis, which tends to exacerbate the heat problem.

Switch mode supplies are not as simple to check as linear supplies. For example, if capacitor C1017 in the GE VCR supply were to dry up, the 5 V would be missing. Then, the power supply would not turn on because the photodiode in the optoisolator is not getting enough current and the phototransistor will not change its value. So the whole power supply becomes a problem. But most of the time, by checking the components, you can locate the one that has failed.

If for some reason you forget to replace the shield of a switch mode power supply, you will create havoc with the video. The switching transformer emits very high frequencies, about 100 kHz, not like a 60 Hz transformer. The video head, which is basically a coil, can easily pick up these frequencies. The high frequency of operation also accounts for the small size of switch mode transformers.

What is the symptom of a failure in the power supply? As mentioned, the strongest indication is the absence of the clock on the display. If you power the VCR and the clock is not blinking, chances are the power supply circuit is dead. You get this same symptom if the microprocessor is dead, but it is much more likely that the power supply is the problem.

Electronics parts suppliers sell kits for servicing switch mode power supplies. The kit includes all the components that go bad, capacitors, diodes, transistors. You can replace the whole supply with these new parts. This way you don't overlook anything. When replacing parts, you must pay close attention to the polarity of the capacitors and diodes. You have to install them properly.

Very rarely do you have to change the switch mode transformer. You need special equipment to check them, such as a Sencore analyzer. Sencore has a patented ringing test, which tells you immediately if the primary of the transformer is good.

Chapter 17

Troubleshooting All-In-One TV/VCRs

All-in-one TV/VCRs are becoming more popular, and thus, you will see more of them come into the shop for repairs. There are no major differences between an all-in-one TV/VCR and two separate units.

The biggest difference is that the VCR in the all-in-one unit does not have an rf converter. In other words, the video and audio signals don't get converted to channel three and then supplied to the tuner of the television. Instead, the composite video and audio signals from the tuner/demodulator go directly to an electronic switch that connects to the video amplifier inside the television. These signals bypass the television tuner and give you a slightly better quality picture. In other words, the signal-to-noise (S/N) ratio is slightly better because there is no rf converter and no TV tuner involved. You bypass two major units that are sources of noise. But, then again, most of the VCRs in all-in-one units are basic 2-head machines, so they do not produce great quality anyway.

Some TV/VCRs, such as the 19-inch Samsung CXA-1926B shown in **Figure 17.1**, use separate power supplies for the television and the VCR. In this case, each one has its own switch mode power supply in a shielded enclosure. Some all-in-one units have one power supply for both TV and VCR. Having two separate supplies makes it easier to work on the VCR, since you can essentially remove it from the TV and work on it by itself.

When you remove the back cover of the Samsung TV/VCR (**Figure 17.2**), you'll notice that the VCR looks like a self-contained unit (**Figure 17.3**). In fact, it has its

Figure 17-1. An all-in-one TV/VCR.

own metal top cover and metal bottom cover. But the VCR has no display. All the VCR functions are available through the on-screen display. Most of the functions are on the remote control, not on the VCR itself. In order to program the VCR, you need the original remote control.

As far as servicing goes, the real inconvenience is that the VCR section has to be removed from the housing of the television (**Figure 17.4**). This is a little hassle but no major feat.

On some models, such as Panasonic all-in-ones, the VCR is at the top of the unit. This type is tougher to work on than all-in-ones with the VCR at the bottom.

For all-in-one units, you obviously have to have some knowledge of TV circuits to fix all the problems that may occur. In many cases, the customer complains that the TV is not working, whether or not it is the source of the problem.

17.1 Determining the Source of the Problem

The VCR in an all-in-one unit works like any normal VCR. Any problem that you may encounter in a 2-head VCR can pop up in the VCR of an all-in-one unit. The idler arm assembly may wear out, the power supply may short out, the video head may get dirty.

Figure 17-2. Removing the back cover of the all-in-one TV/VCR.

Figure 17-3.
The internal VCR has metal top and bottom covers.

Figure 17-4.
Removing the VCR section from the housing of the television.

Figure 17-5.
The VCR with the top cover removed for servicing.

It's not that difficult to determine where the problem lies once you open the top cover of the VCR (**Figure 17.5**). You'll be able to see if all the mechanical functions are working properly. If the VCR is working mechanically, but the power doesn't stay on, then you have to examine the TV power supply. Disconnect the VCR altogether. **Figure 17.6** shows the connector that goes to the video circuits of the TV. Then, check if the TV power supply is working. If it's not, then you must repair the TV power supply and reconnect the VCR. If the TV is working, then you have to repair the power supply in the VCR.

If any belts are broken or if you need to access the chassis from the bottom of the machine for any reason, there is sufficient play in the wires to turn the VCR on its side as shown in **Figure 17.7**.

17.2 Safety Considerations

There are additional safety considerations when you are working with any television. A 19-inch TV has a high voltage of about 22,000 volts. The highest voltage in a VCR is the one that comes from the AC line—120 volts. You have to stay away from the picture tube in a TV, because this is where the high voltage is (**Figure 17.8**). Again, the best solution is take the VCR completely out of the TV and do the repair. If you have a problem with the TV, then you have no choice but to repair it. You really need the expertise to work on both; otherwise, you shouldn't start.

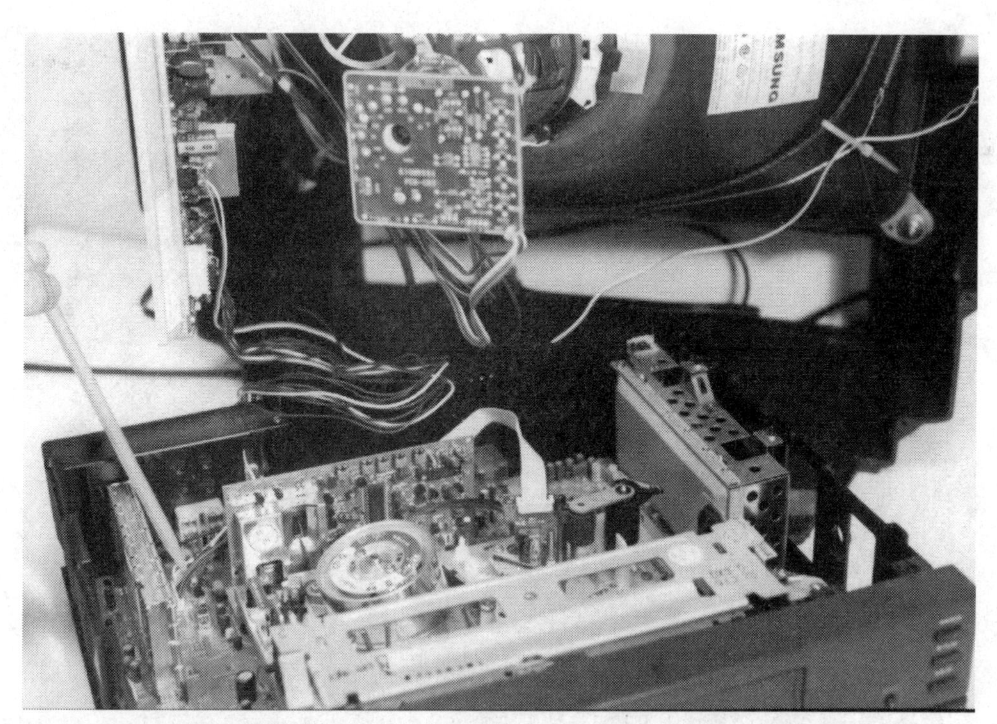

Figure 17-6. The connector in the VCR that goes to the video circuits of the TV.

Figure 17-7. Turning the VCR on its side to remove the bottom cover is not a problem.

Figure 17-8. The high voltage comes from the picture tube of the TV, a dangerous area.

Appendix A

Case Study 1: Fisher Model FVH904A

This VCR had no ability to rewind or fast forward. When you pressed fast forward, you could hear the motor spinning, but the tape didn't move and a few seconds later, the power shut off.

These symptoms point to a mechanical problem because the power supply, the system control, the cassette loading motor, and the capstan motor all appear to be working. Remember, it is the microprocessor that is shutting the power off since it has sensed a problem.

We removed the top and bottom covers to examine the mechanical parts of the VCR. On this model, when the bottom cover is removed you can see all the gears, the belts, and the capstan motor.

When we pressed fast forward we could see the capstan spinning and all the wheels moving. The only part between the reel disk and the wheel driven by the belt is the idler arm assembly—a wheel with a rubber tire. On the bottom of the wheel there is a felt clutch assembly. The purpose of the clutch assembly depends on the direction of the wheel. The clutch slightly resists the movement and goes left for rewind and right for fast forward. If the clutch is not there the wheel will not move.

The idler arm assembly seemed to be causing the problem. To check for sure, the easiest thing to do is put the test jig in the VCR and activate fast forward. With one hand holding the right take-up reel, you can see if the idler assembly is slipping. It should transfer the torque from the wheel driven by the capstan motor to the take-up reel. It didn't.

Removing the idler arm assembly in this VCR requires a lot of disassembly. We began by releasing the printed circuit board which has four screws, one attached to the cassette housing. We also needed to release a plastic clip. Then, we lifted the PC board up and out of the way.

Next, we removed four screws from the cassette assembly and disconnected a plug that goes to the main circuit board. Then we lifted out the cassette housing. Next, we removed a plastic c-clip from the top of the take-up reel and pulled out the reel. Then, we removed a spring from the clutch assembly (part of the idler assembly) to the center shaft.

At this point, we turned the VCR upside down and removed the large belt from the wheel that turns the take-up reel. Then we removed the c-clip from the wheel and pulled the wheel out of the VCR. The only thing left to remove was the c-clip from the arm of the idler assembly. We removed this and pulled the idler assembly out. Then we removed the tire and replaced it with a new one. Finally, we reassembled the machine doing everything above in the reverse order. In an older VCR like this one, it's a good idea to perform routine maintenance such as cleaning and lubrication. Figures A.1 through A.18 show how the repair was done.

Figure A-1. The Fisher FVH904A VCR.

Figure A-2. Unscrewing the top cover.

Figure A-3. Lifting off the top cover.

Figure A-4. Unscrewing the bottom cover.

Figure A-5. Taking off the bottom cover to observe the VCR mechanisms.

Figure A-6. Unclipping the front panel.

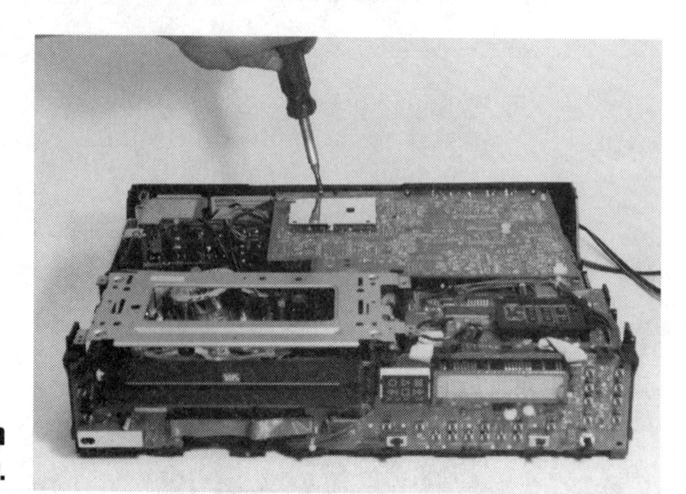

Figure A-7. Unscrewing the main circuit board to free the cassette housing.

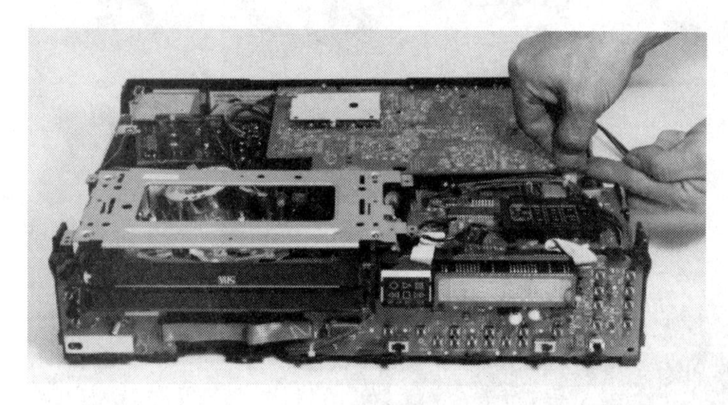

Figure A-8. Unclipping the main circuit board.

Figure A-9. Unscrewing the cassette housing.

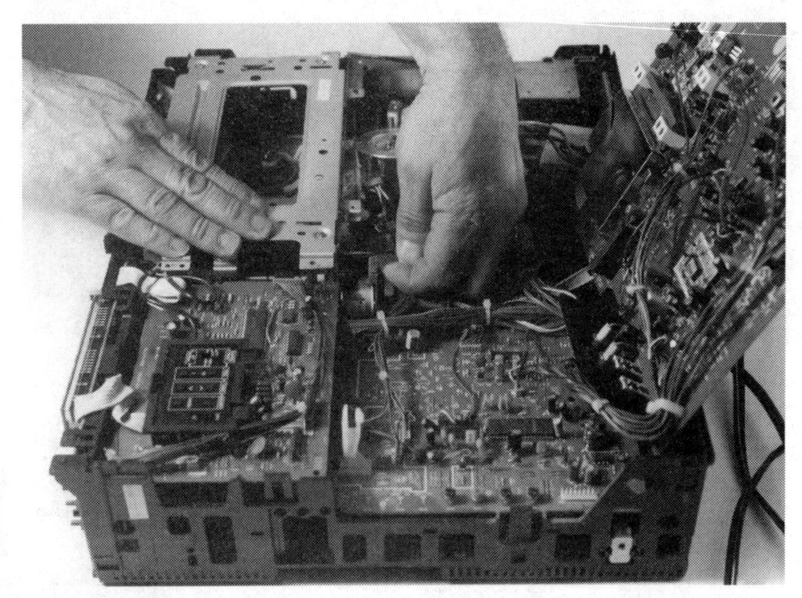

Figure A-10. Removing the electrical connector from the cassette housing.

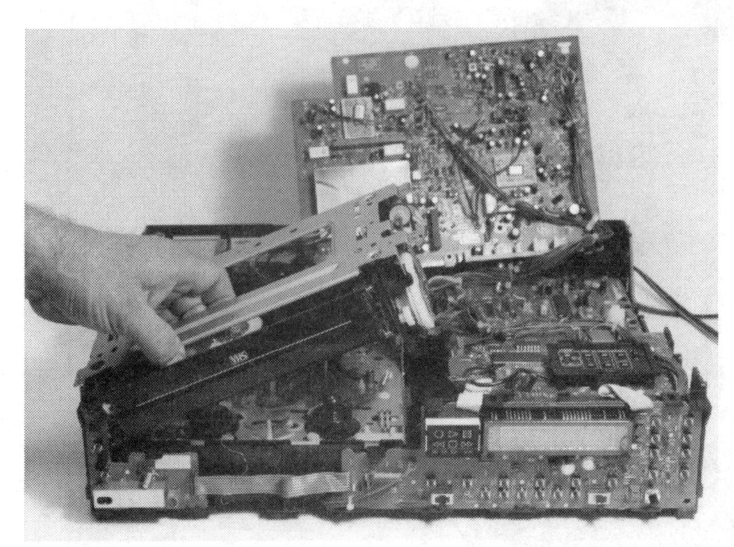

Figure A-11. Lifting the cassette housing out of the VCR.

Figure A-12. Lifting off the takeup reel.

Figure A-13. Using a special hook tool to remove a spring from the idler arm assembly.

Figure A-14. Pushing part of the idler arm assembly through a hole in the chassis.

Figure A-15. With the VCR now upside down, removing the clutch.

Figure A-16. Removing the idler arm assembly.

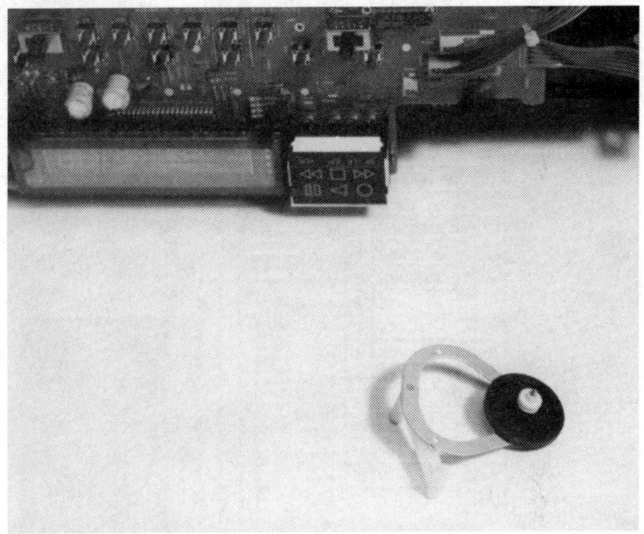

Figure A-17. The idler arm assembly has a worn tire.

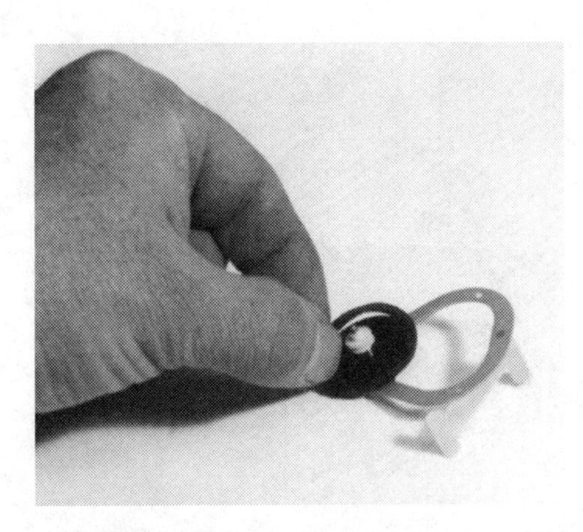

Figure A-18. Removing the worn tire from the idler arm wheel. Replacing this tire fixed the problem.

Appendix B

Case Study 2: Magnavox Model VR9845AT

This VCR would not play or perform any other tape functions. When you pressed PLAY, the icon indicator would come on but nothing happened. This could be a mechanical or an electrical problem. It is really not clear from the symptoms. The power supply appears to be working, since the indicator light comes on.

We removed the bottom cover from the VCR to examine the mechanical components. This VCR uses the capstan motor for loading the tape through a series of gears. There is not a separate loading motor. This makes it potentially complicated to perform the repair.

We pressed PLAY and noticed that nothing moved inside the machine. The capstan motor did not spin. This was our first clue to the cause of the problem. This type of problem, where the capstan motor does not even start to spin, is often caused by a problem with the mode switch. Of course, this is only true if the power supply is working properly, which seemed to be the case.

We removed the top cover of the VCR and examined the position of the mode switch. The mode switch is located underneath the printed circuit board, directly on the chassis. The mode switch has a dot, which has to be aligned with the mark on the base of the switch with the tape ejected from the VCR.

Usually, when you put the tape in a working machine, the mode switch dot moves clockwise about one-quarter turn. If this happens, the mode switch is set correctly. In this case, the mode switch was a distance of one gear tooth off position in a counterclockwise direction.

This problem occurs because sometimes people force the tape into the machine, which makes the gears jump out of alignment. We decided to place the mode switch in the right spot to see what would happen. To do this we needed to disassemble the VCR.

We removed the front panel of the VCR and then removed the cassette housing. Next, we lifted up the hinged printed circuit board about 45 degrees and angled a piece of plastic against it to prevent the board from falling down. Then, we removed a c-clip on the top of the pinch roller with a small screwdriver. Once that clip is removed, the whole pinch roller assembly can be lifted out easily. But before you do this, you should make a mark or a drawing to show how the pieces are aligned. Later on, when you put it back together, you will not have any problems.

After removing the pinch roller assembly, we lifted out the cam gear. Then, we pulled the mode switch up slightly so that we could turn it one notch down to the correct spot. We watched carefully as we put the cam gear and pinch roller back in place to make sure these parts didn't change the position of the mode switch.

After we put the pinch roller assembly back in place, we had to put the cassette housing back in place and secure it with the screws. Then, we tried to play the tape. If we did everything properly, the VCR would work. In this case, fast forward and rewind worked, but PLAY did not. We had to repeat the disassembly process and re-align the gears again. Then, we reassembled everything and gave it another try. This time, all the VCR functions worked properly.

As you can see, proper alignment is imperative for the VCR to operate. Aligning the mode switch is tricky. If you forget to mark the alignment, you can try to align the gears by trial and error. If this doesn't work, you will have to obtain the service manual from the manufacturer, which provides the alignment procedure. Figures B.1 to B.9 show how the repair was done.

Figure B-1. The Magnavox VR9845AT VCR.

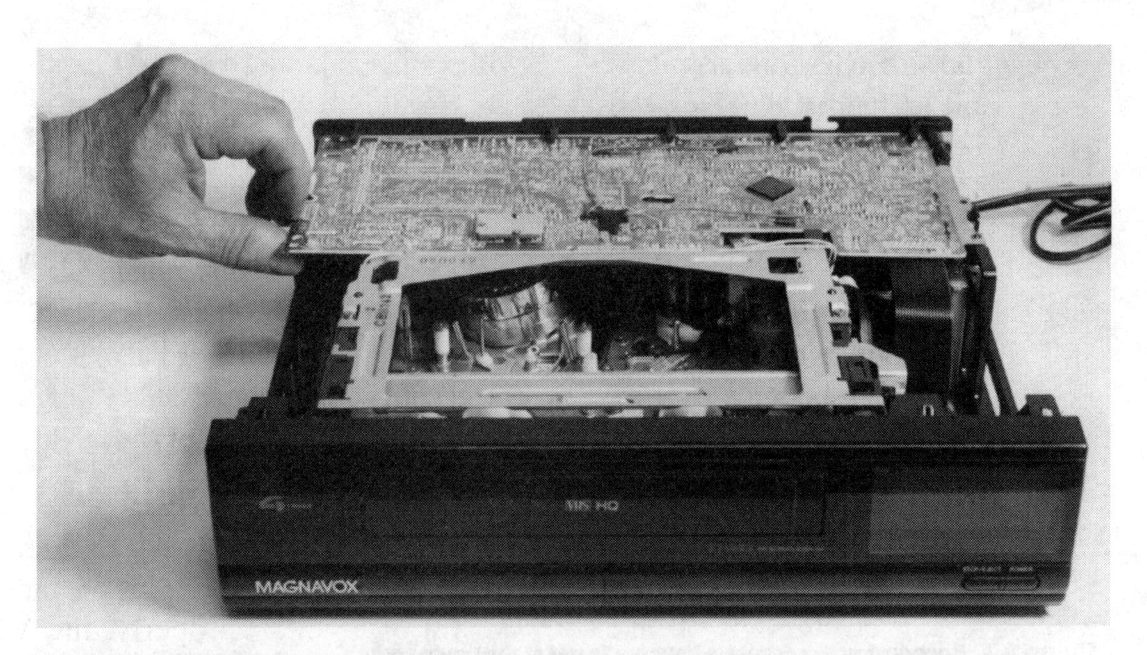

Figure B-2. With the top cover removed, unclipping the main circuit board to take a look at the mode switch.

Figure B-3. We observed that the mode switch was off by one notch of the gear.

Figure B-4. Unscrewing the cassette housing to get at the mode switch.

Figure B-5. Lifting out the cassette housing.

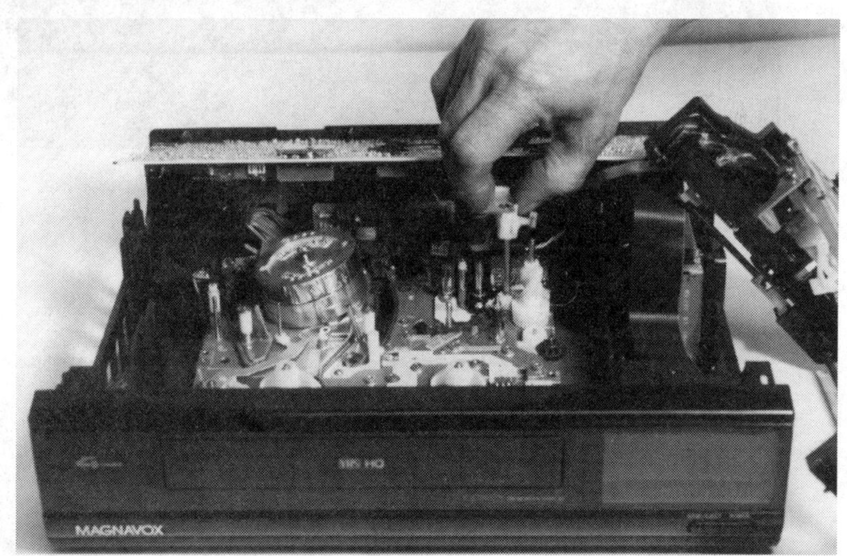

Figure B-6. Lifting out the pinch roller.

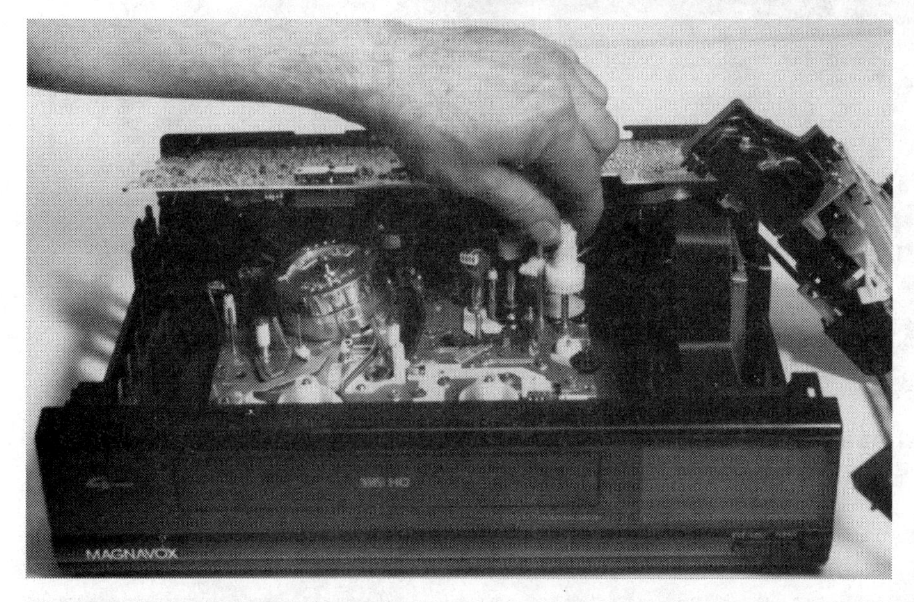

Figure B-7. Lifting out the cam gear.

Figure B-8. Moving the mode switch one notch with a screwdriver.

Figure B-9. A closeup of the mode switch showing the dot used for alignment. Note the switch is mounted on a small PC board with five connectors. Realigning the mode switch solved the problem.

Appendix C

Case Study 3: GE Model 1VCR5011X

This VCR had a problem with timed recording, according to the owner of the machine. When the preset time arrived, the VCR would begin to record. Then, in a few seconds it would stop, then it would start again, then stop, then start, and on and on. While this was happening, the timer icon and the REC label would blink on and off.

This appeared to be an intermittent electrical problem, one that you often see when there is a loose connection or a cracked solder joint. We removed the top cover of the VCR and the front panel. To check if there was a loose connection, we tapped the printed circuit board with the plastic handle of a screwdriver in the area of the timer IC. We did this while the VCR was in the timed recording mode. We could not force the problem to occur.

Thinking that this problem might be caused by the power supply, we checked the voltage going to the timer IC. We needed the service manual to do this. The timer chip, a UPD7538, has voltages on pins 21 (+5 V) and 4 (-3.6 V). Both of these voltages were good and stable, even while the VCR was turning on and off.

When checking an IC like the timer chip, you have to make sure that the voltages are powering the chip and also that the crystal is operating. We checked the frequency of the crystal with the frequency counter on our Sencore Universal Video Analyzer. This is convenient for us to do, but this kind of measurement can be made with any scope or frequency counter. We measured about 401 kHz, which is the correct frequency (marked right on the crystal). Again, this waveform was stable even while the VCR was performing intermittently.

Every other part of the VCR seemed to be working except timed recording. Since the voltages and clock frequency checked out good, we wondered if the timer IC was bad. Replacing this IC is no easy job. We measured the voltage at pin 6 of the timer IC, the pin that goes to the timer icon on the front panel display. This voltage jumped up and down whenever the VCR displayed the intermittent behavior. We couldn't tell for certain, though, if the timer IC was causing this.

There are many sensors and switches that turn voltages on and off in a VCR. These signals are sent to the microprocessor. The microprocessor then responds by turning various IC pins on and off, including many pins on the timer IC. We decided to check one of these switches, the safety tab switch. The purpose of this switch is to prevent the VCR from recording when the tab on the tape cassette is broken. This protects against inadvertently taping over a prerecorded tape. For normal tapes, the switch is pushed down on a hinge. This closes the switch and creates a direct path to ground since one of the switch's pins are grounded to a screw on the chassis. When the switch closes, a low signal is sent to the microprocessor and the VCR is allowed to record. When there is a broken tab or no tape at all, the switch stays open. This sends the microprocessor a high signal and recording is not allowed. In the meantime, the microprocessor signals the timer IC to turn the timed recording signal on and off, if the VCR happens to be in the timed recording mode.

We used a DMM in the ohmmeter mode to measure the switch as it opened and closed. The reading was erratic. We sprayed the switch with WD-40 and measured again. The reading was still erratic, which meant the switch was still not making a good contact when closed. Then, with a small screwdriver, we gently scraped the contact to remove the oxide or whatever dirt was causing the problem.

Now, when we measured the closed switch, the ohmmeter showed zero ohms. This indicated that the switch was working properly again. Now when we tried timed recording, the intermittent problem was gone.

The original complaint from the owner concerned timed recording. But the safety tab switch works for manual recording, too. During the course of our testing, the manual recording seemed to work okay. We must not have left it on long enough for the problem to occur. After all, this was an intermittent problem. Figures C.1 to C.7 show how the repair was done.

Figure C-1. The GE 1VCR5011X VCR.

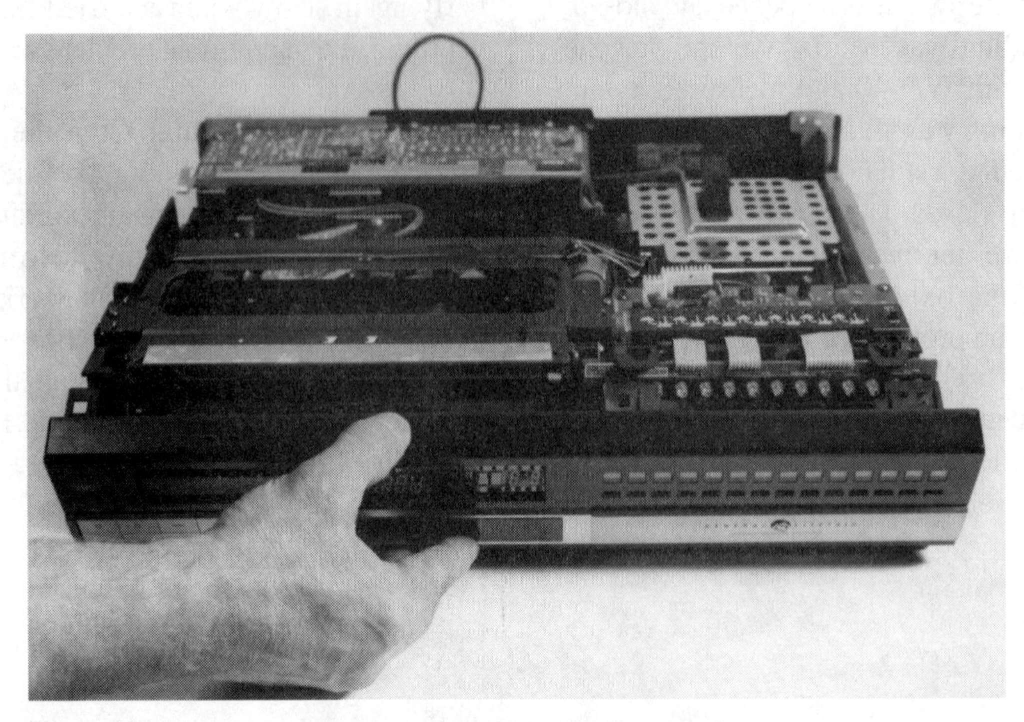

Figure C-2. With the top cover removed, detaching the front panel.

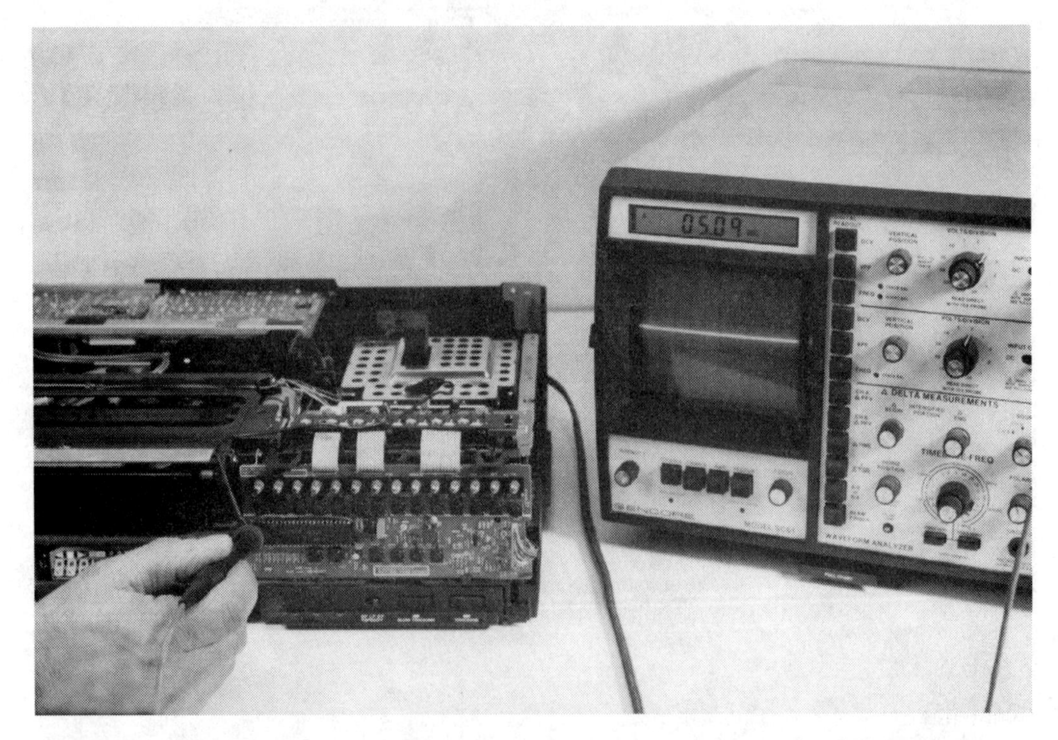

Figure C-3. Measuring the VDD voltage on the timer chip using the Sencore SC61 Waveform Analyzer.

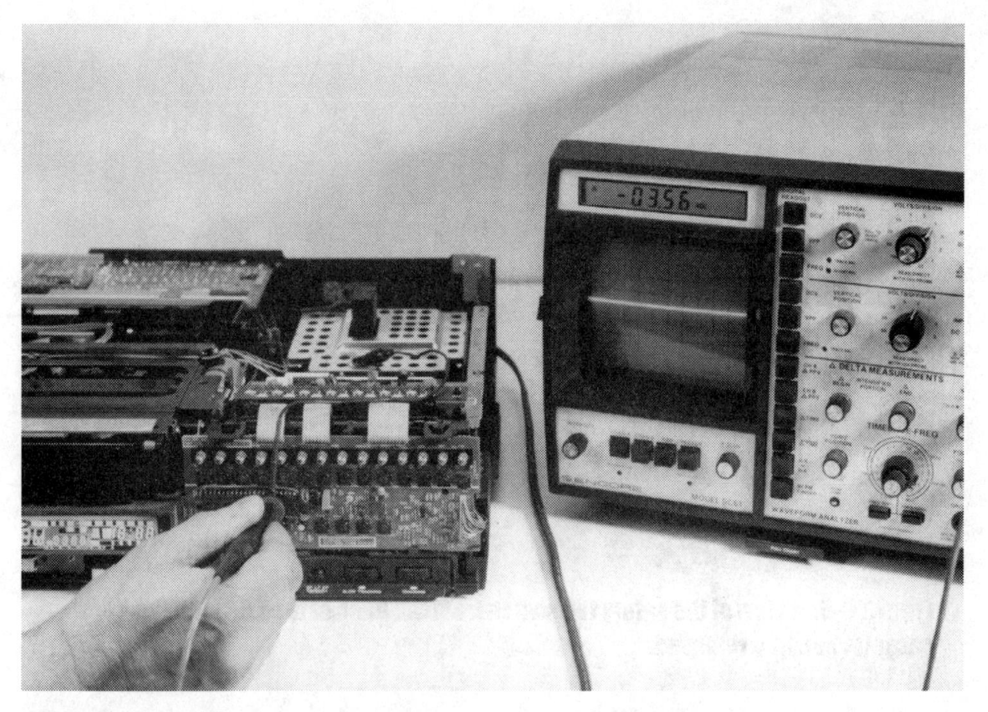

Figure C-4. Measuring a negative voltage with the Sencore SC61.

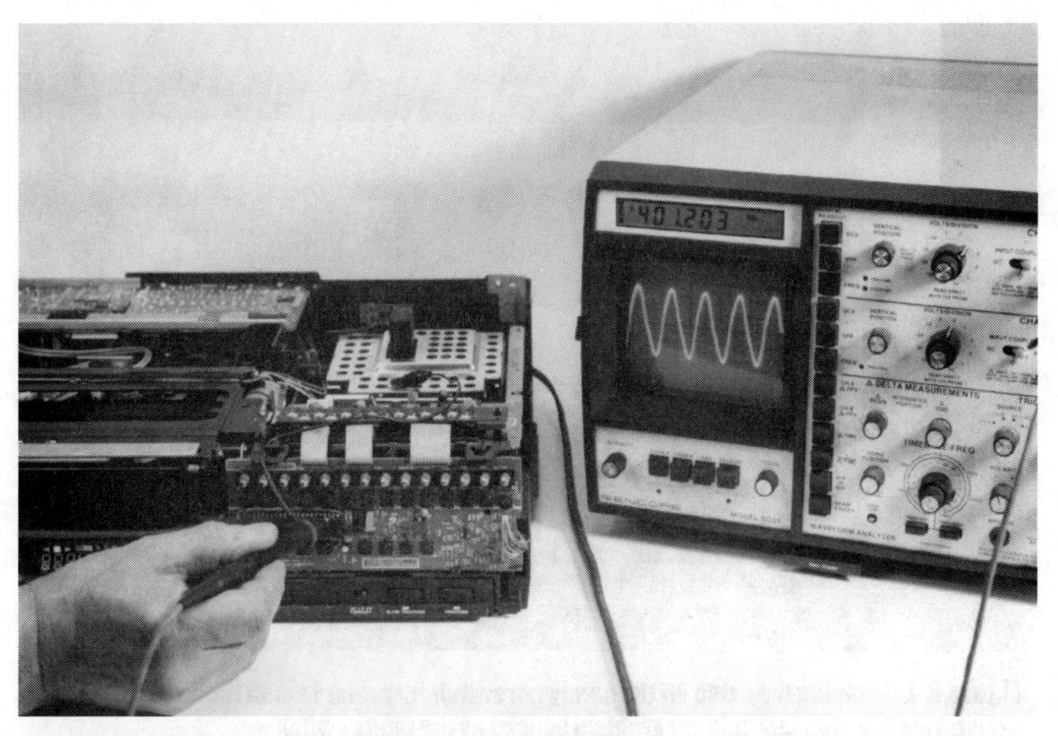

Figure C-5. Measuring the frequency of the timer crystal with the Sencore SC61. The value appears just above the sine wave displayed on the scope.

Figure C-6. A view of the safety tab switch from the back of the VCR with the cassette housing removed.

Figure C-7. Scraping the oxide off the safety tab switch with a very small flathead screwdriver. This fixed the intermittent problem caused by the faulty switch.

Appendix D

Case Study 4: RCA Model VJT275

This VCR had a tape-in LED that stayed on even though a cassette tape was not loaded. If you tried to load a cassette, it wouldn't go into the machine. The lit tape-in light points to an electrical problem.

The tape-in light is activated by a switch located under the loading motor. To check this switch you have to remove the cassette housing, which takes some time. A component that activates the loading motor is the left tape-end sensor. Rather than remove the cassette housing to check the switch, we decided to first check the sensor.

We measured the voltage at the tape-end sensor when it is exposed to the light from the light emitting diode (LED) in the center of the loading area. You have to measure the voltage at the emitter of the sensor. This is marked on the small PC board that holds the sensor. The voltage is equal to the collector voltage, which is about 9.5 VDC. When you cover the LED, the voltage should drop significantly, to about 3 V. If you cover the sensor itself so that no light comes in at all, then the voltage will drop to zero. If this happens, the sensor is working properly.

In our case, the voltage did not change at all when we covered the light from the LED. By the way, we covered this LED with a small rubber cylinder, to make it easy to take the measurement. Obviously, the sensor was not working at all.

To replace the sensor, you must desolder it from the small PC board. You need to pay particular attention to the orientation of the sensor in the PC board so that you can install the new one correctly. Sensors can be purchased from any good electronic supply stores. There are many different kinds of sensors. The one we needed was an Hitachi part number 5380132. Why an Hitachi part for an RCA VCR? Because Hitachi is the manufacturer of this VCR even though it is sold under the RCA brand name. The best thing to do if you are not sure about the part, is to look in a catalog or ask the supplier for the correct replacement part. Otherwise, you would need to look up the part in the service manual.

After we installed the sensor, we checked the voltages again. When we covered the sensor so that no light could get in, and we measured about 0 V. This indicated that the sensor was now working. We tried to insert a tape. This time the tape-in LED was off. The VCR took the tape without a problem. We made sure to check that the tape-end sensor was working properly by fast forwarding to the end of the tape and making sure that the tape changed direction automatically when it reached the end of the tape. It did.

Since this machine was fairly old, we performed routine maintenance, checked the belts, cleaned the heads, and lubricated the mechanisms. This assures us that the owner will not experience any problems due to poor maintenance in the near future. Figures D.1 to D.7 show how the repair was done.

Figure D-1. The RCA VJT275 VCR with its top cover removed.

Figure D-2. The cassette-in LED was on even though the cassette tape was not loaded into the machine.

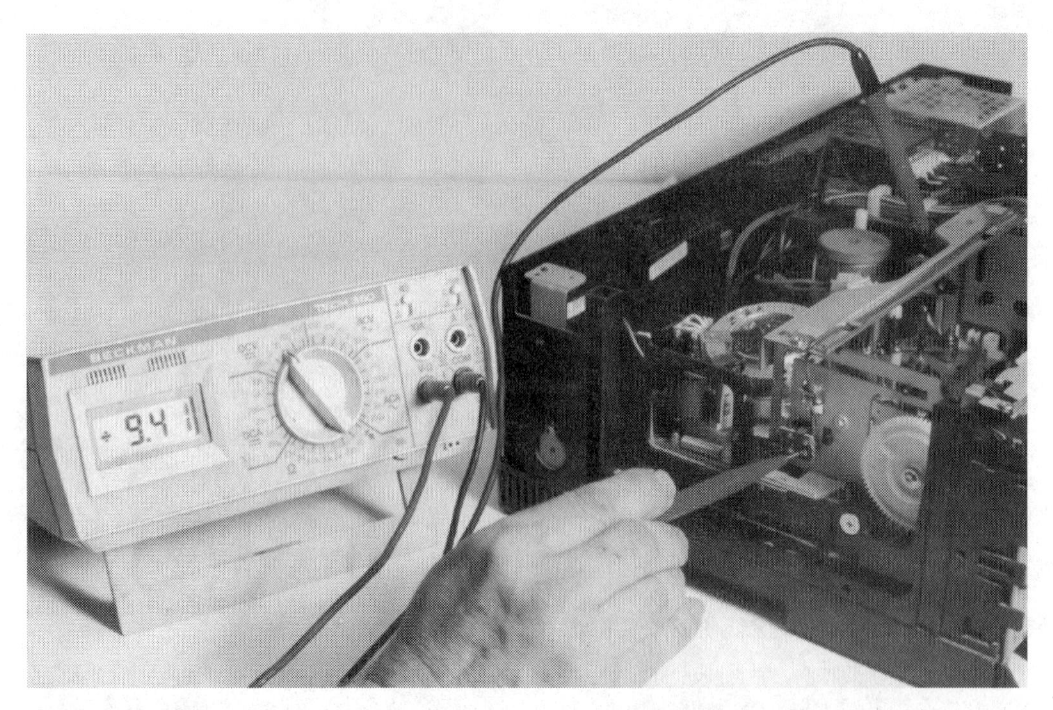

Figure D-3. Measuring one of the tape-end sensors. The high 9.41V reading didn't change even when we blocked the light to the sensor.

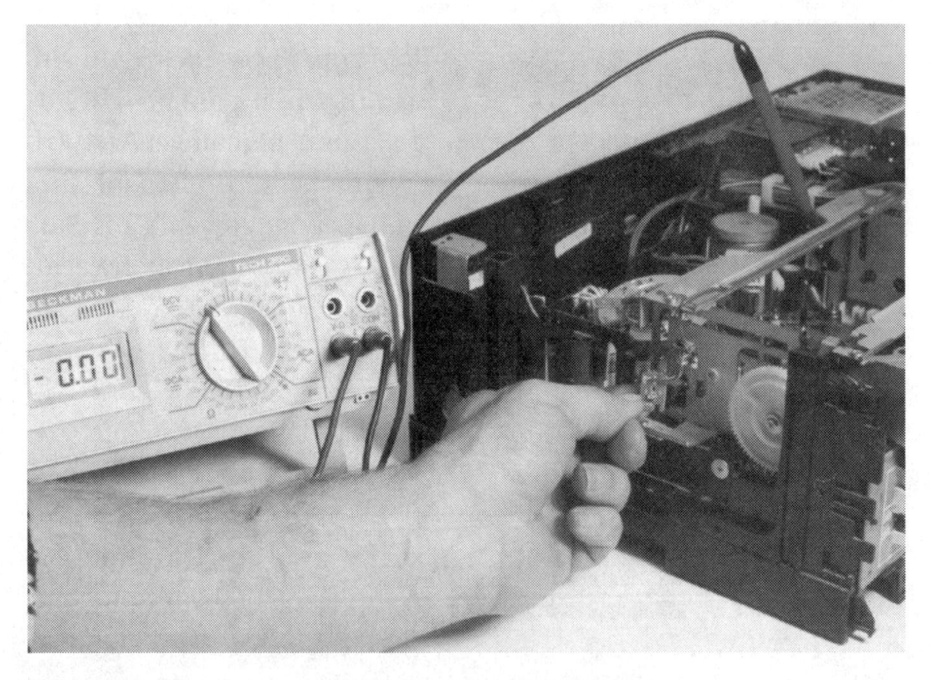

Figure D-4. The sensor is mounted on a very small PC board.

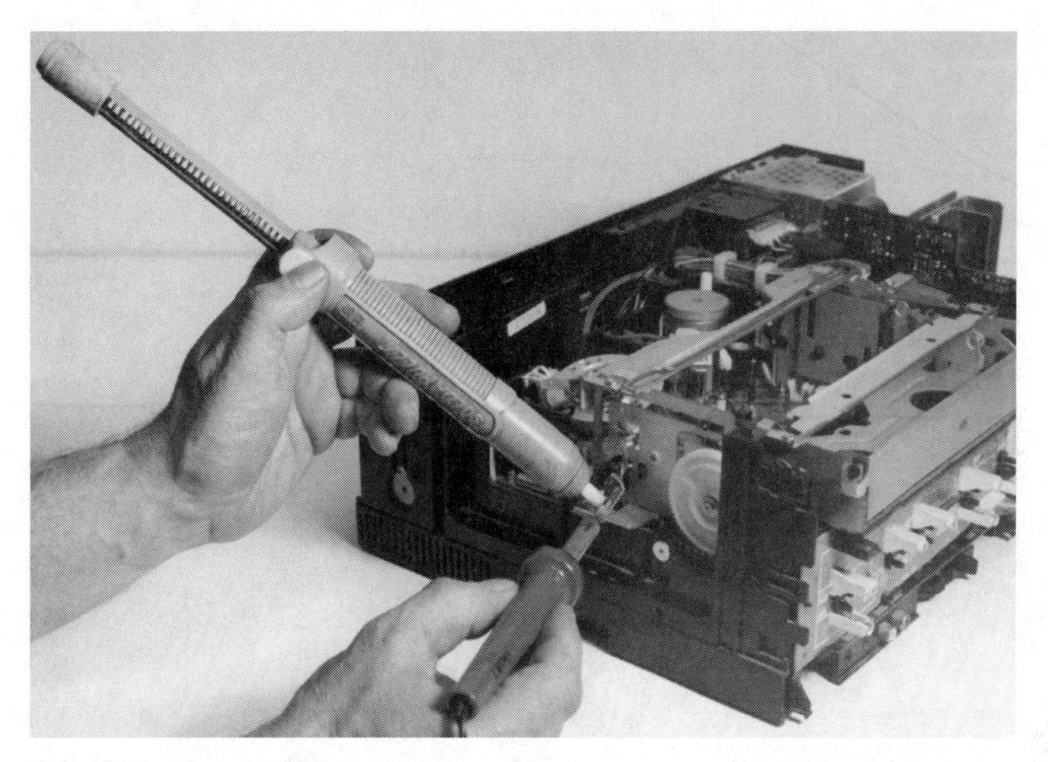

Figure D-5. Desoldering the sensor.

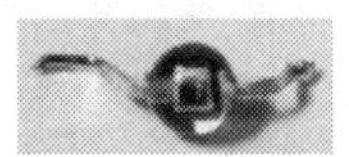

Figure D-6. A closeup view of the sensor, a photo transistor.

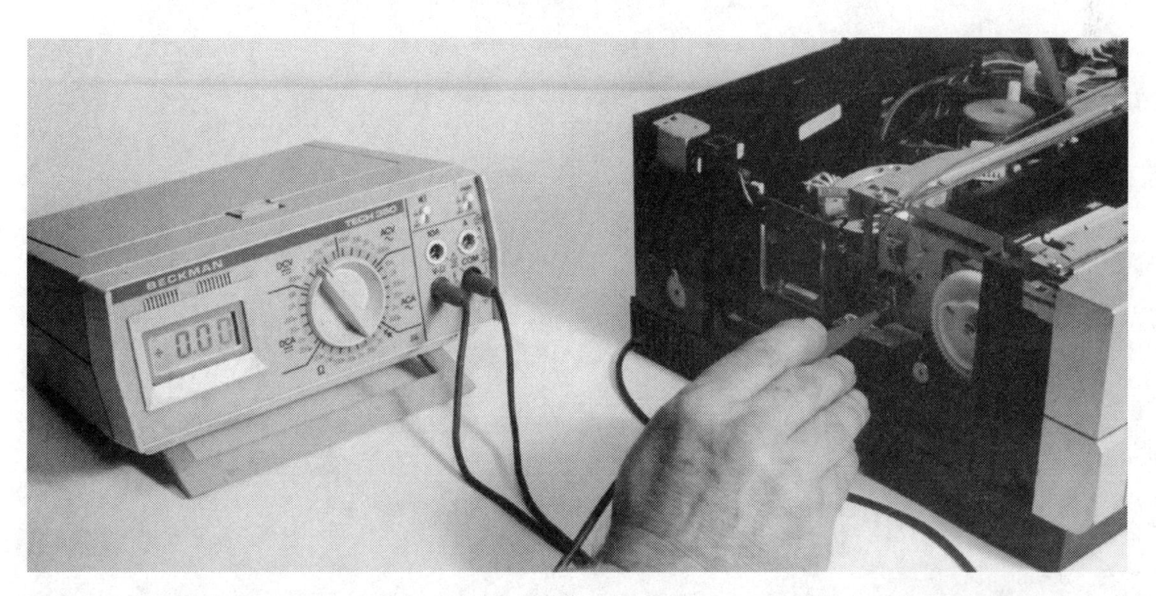

Figure D-7. With the new sensor in place, the voltage goes down to zero when we block the light.

Appendix E

Case Study 5: RCA Model VR285

This VCR had a tape loading problem; it would not accept the tape. Also, the VCR was making a whirring noise. These symptoms point to a mechanical problem with the VCR, which you can observe best in this VCR by removing the bottom cover.

We removed the bottom cover and, after close inspection, noticed a small wheel with a broken belt beside it. This obviously was causing the problem. In this particular model, replacing this small belt is a big job. The whole loading assembly must be removed. If you do not have experience doing this type of disassembly, you must take special care to note how all of the gears, brackets, springs and clips are oriented. If you don't do this, you risk not being able to reassemble the VCR.

The trick to disassembly is to remove a c-clip from one of the brackets that moves the pinch roller. To get at the shaft of this piece, we had to remove the top cover of the VCR. Once the c-clip is removed, the next step is to remove the three screws that hold down the loading motor and gears. After removing the screws, we had to maneuver the assembly so that a pin in one of the metal brackets could be removed from a hole in another metal bracket, which is just under the front cover. We noticed a small spring in this area and so made a mental note to be careful not to dislodge and possibly lose it. Whenever you do a job like this, you must pay close attention to how pieces fit together so that you will know how to reassemble them.

After removing the loading assembly, we turned it upside down to see the belts and pulley wheels. There are two belts, one is about twice the size of the other. The small belt must be removed first. Remember, though, the larger belt had snapped, so only the small belt was there.

To install the large belt, we used a small screwdriver to unclip the entire shaft that includes the worm gear. You have to be very careful removing these clips because there are small washers between the clip and the gear.

Once we removed this piece, we placed the belt over the motor pulley first, then over the worm gear pulley. Next, we stretched the belt slightly and put the clips back in place. Then, we reinstalled the small belt.

The final step is to carefully clean the belts with cotton swabs and alcohol to remove any grease that may have gotten on the belts during installation. While installing the belts, we made sure not to move any gears that would affect the alignment of the VCR.

We had to be very careful putting the loading assembly back in place. We made sure the two metal brackets seated properly together with the pin in the hole. We also checked that the spring was in place. Then, we had to check that the tension spring on the left gear was over the gear, not under it. When we were sure everything was correctly in place, we gently pushed down the loading assembly.

Three legs have to fit right onto the chassis. Then, we screwed the assembly down. Finally, we turned the VCR rightside up and placed another bracket over the shaft and put the c-clip into place. Figures E.1 to E.10 show how the repair was done.

Figure E-1. The RCA VR285 VCR.

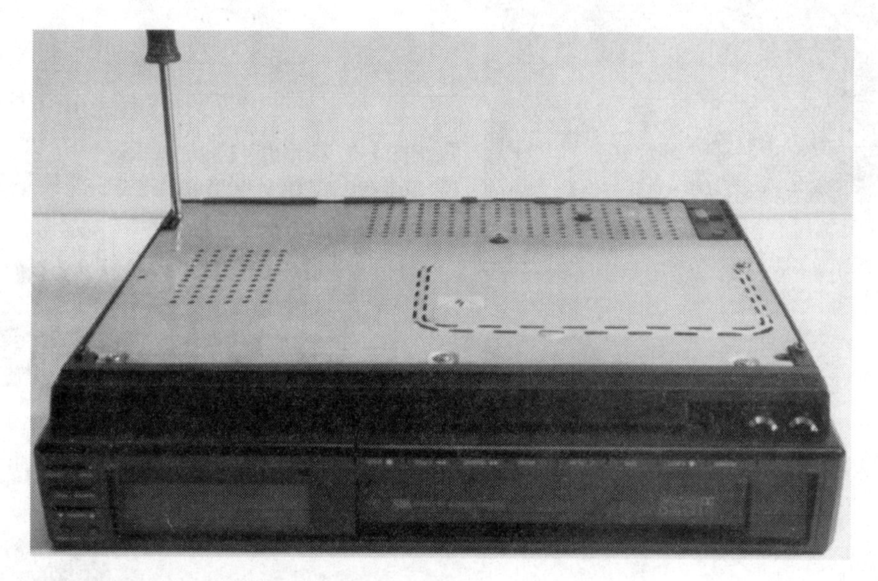

Figure E-2. Unscrewing the bottom cover.

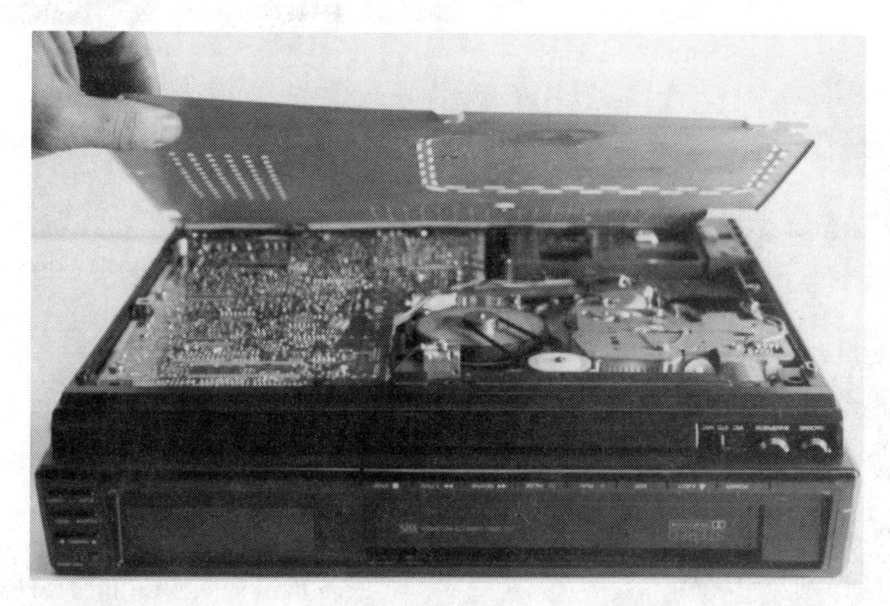

Figure E-3. Removing the bottom cover to get to the loading wheels and gears.

Figure E-4. Through observation, we noticed a wheel without a belt.

Figure E-5. Removing the loading mechanism to get to the wheel.

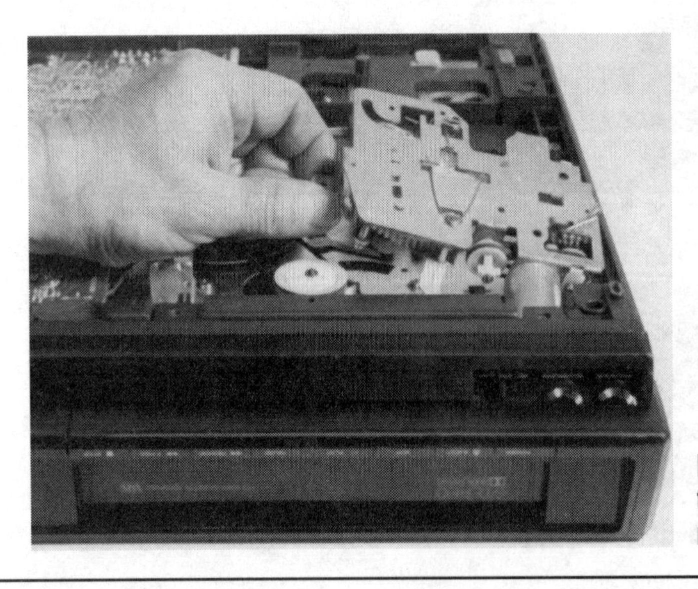

Figure E-6. After removing three screws, we lifted the loading mechanism out of the machine.

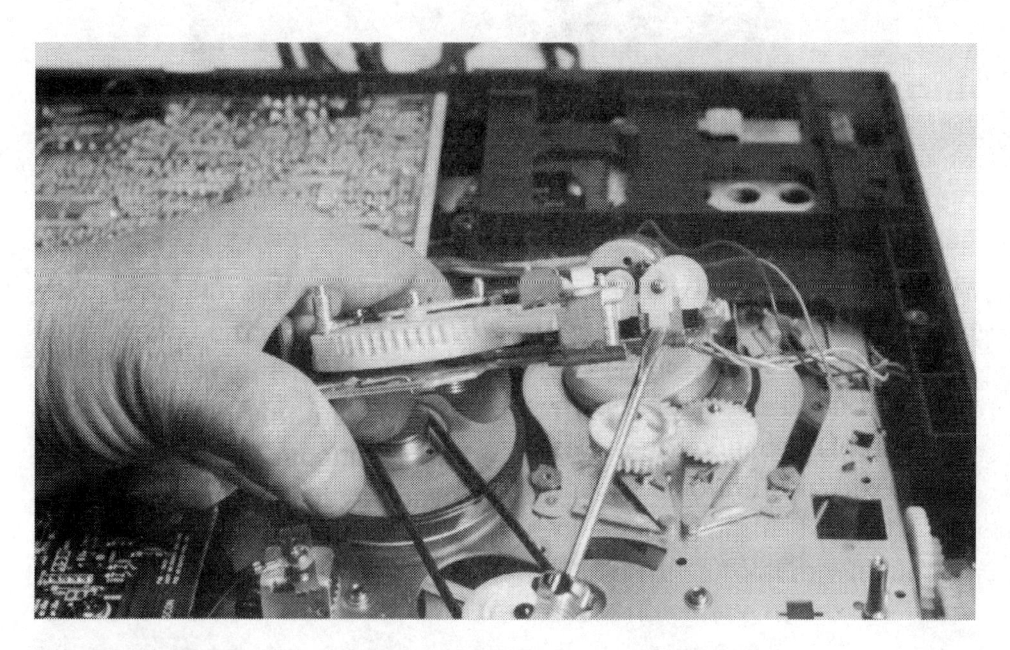

Figure E-7. Prying up the worm gear shaft to get to the motor pulley.

Figure E-8. Removing the worm gear shaft.

Figure E-9. The belt is now back on the motor pulley.

Figure E-10. Checking the tension spring on the left gear to make sure it was over the gear, not under it. Once the reassembly process was complete, the VCR worked perfectly again.

Appendix F

Case Study 6: RCA Model VPT293

This VCR would play for a about a minute and then shut off. This is a popular problem with this model.

We placed a test jig in the machine to find out the cause of the problem. When we pressed PLAY, we saw that the right reel started turning slowly and a few seconds later stopped. Then, the VCR shut off.

We took off the bottom cover to check the belts. This machine has two belts for the loading assembly, one long square belt from the capstan motor to the idler arm assembly and another small one from the idler assembly to the clutch. We checked the belts; they were okay. We pressed play again and watched how the reels turned from the bottom of the machine. We saw that the clutch stopped turning at one point. When we tried to move the clutch by hand, it offered heavy resistance, which should not be. This mechanism should move easily. We looked again at the top of the machine and released the c-clip from the take-up reel and pulled it out. Underneath the reel was a belt that was twisted up. It was acting as a brake. The original purpose of that belt was to turn a wheel on the right side of the take-up reel for a counter. But this model VCR has been modified since then and that belt is no longer needed. Now, a photointerrupter is mounted underneath the take-up reel to do the same job. All that was needed to fix the problem was to remove the belt debris. A new belt was not necessary. The manufacturer blindly adds this belt to this VCR, even though it no longer serves a purpose. Figures F.1 to F.4 show how this repair was done.

Figure F-1. Placing the test jig in the RCA VPT293 VCR.

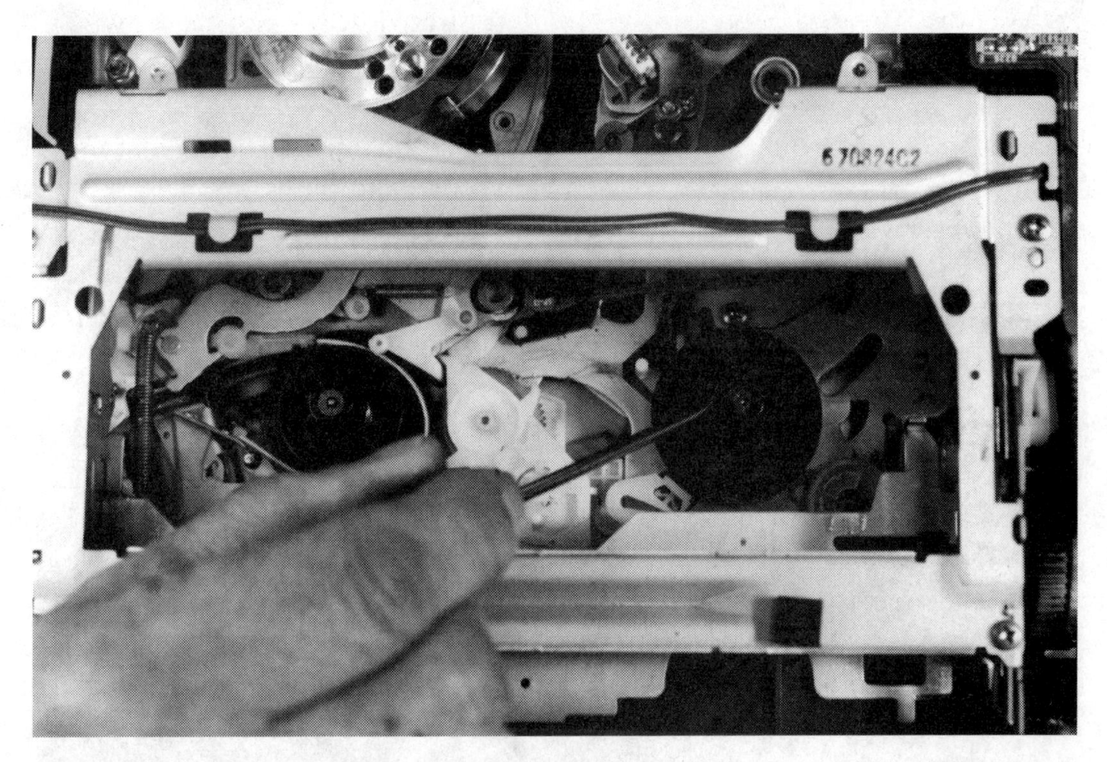

Figure F-2. Removing the c-clip from the take-up reel.

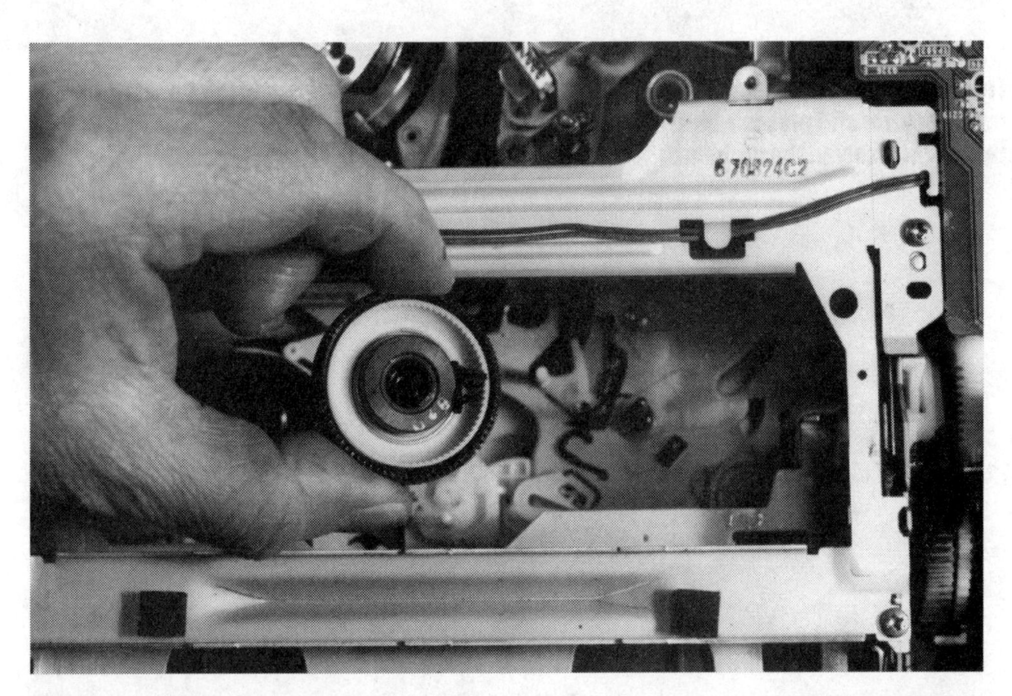

Figure F-3. Examining the underside of the take-up reel, we noticed a belt had broken and wound around the shaft of the reel.

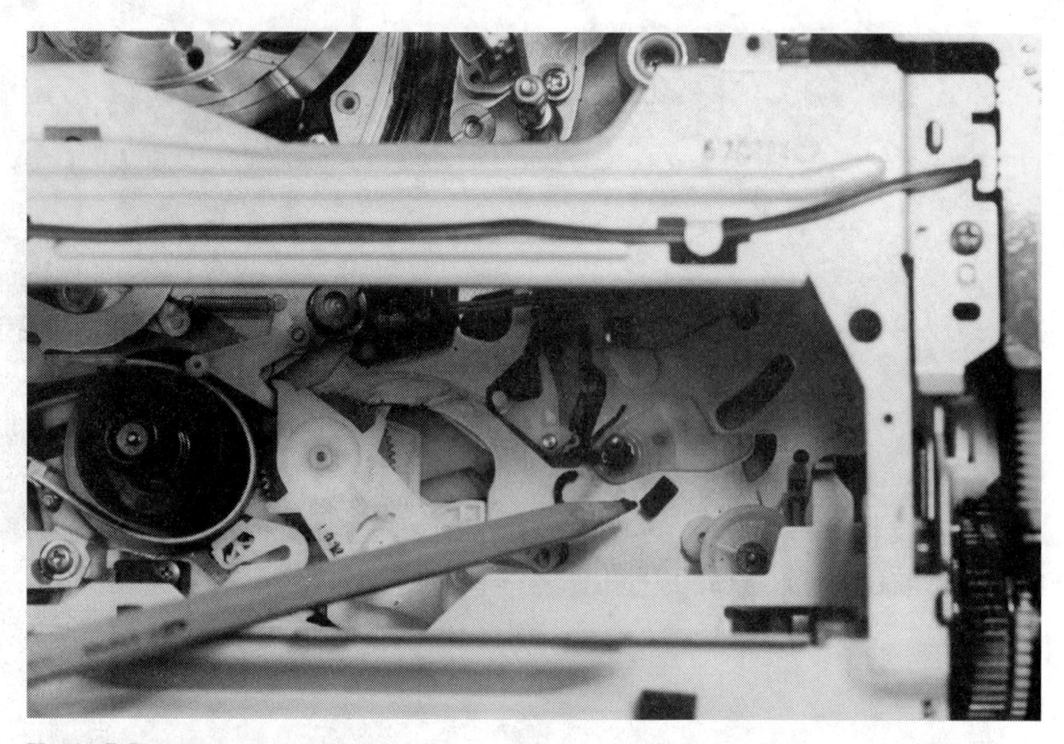

Figure F-4. The new design of this VCR uses a sensor. The old pulley is to the lower right. The manufacturer still places a belt on this wheel even though it is not needed. Removing the twisted belt solved the problem.

Appendix G

Case Study 7: Mitsubishi Model HS-U59

This VCR would play for four or five seconds and then stop. If you tried fast forward or rewind, the same thing would happen. The tape didn't get damaged, but the machine would not play.

We examined the display on the machine and noticed immediately that the counter, which counts seconds and minutes, was not moving. This indicates that the counter sensor is probably bad. This is a common problem with VCRs.

We removed the bottom cover, because normally the leads of the sensor are underneath the chassis soldered to a printed circuit board. This cover had nine screws as well as two plastic bumper feet that had to be removed. Once we had the bottom cover off, we had to locate the sensor. We looked in the general direction of the shaft of the take-up reel. We realized that we would have to remove a gear to get at the sensor. The gear is spring loaded, so with a small flat screwdriver, we pushed the spring to the side and the gear came off its shaft. This gear can be put back in any way, you don't have to worry about alignment.

With the gear off, we could see the four leads of the sensor. We took measurements of the sensor and noticed that the voltage did not change even though we turned the reel by hand. We were expecting the voltage to change from zero to about 1.2 V, just as happens with the supply reel, which has the same sensor. But the voltage did not change. This seemed to indicate a defective sensor.

Next we removed the top cover of the VCR. This VCR has a jog shuttle control on the front panel. We removed the jog shuttle knobs (inner and outer knobs) just by pulling on them. Then we removed the front panel by releasing it from three plastic clips.

Now, we had to remove the cassette housing. This came out easily after we removed four screws. Next, we had to release a tape tension lever. This consists of removing a small spring and c-clip with a small flathead screwdriver. Now we were able to remove the take-up reel and turn it over. A pie shaped pattern of four "slices," alternately black and a shiny surface, is on the bottom of the reel. We inspected the bottom surface of the reel to make sure it was clean.

We now had a clear view of the sensor, a photointerrupter, which was mounted inside a plastic socket. We were fairly certain that the sensor was the cause of the problem. We turned over the VCR, desoldered the leads of the sensor, and replaced it with a similar looking sensor that we had on hand, also a Mitsubishi part. However, this sensor did not work. We were forced to order an exact replacement part from Mitsubishi. This sensor costs about $5.00 and $5.00 for shipping.

We installed the sensor and soldered it into place, making sure that it was seated and aligned properly. There is a notch in the corner of the sensor that lines up with the plastic socket. We soldered the four leads carefully and inspected them afterward for solder bridges.

To check the sensor, the machine has to be in play mode. This means we had to put the cassette housing back in place. This housing came out easily, but was not easy to put back. We had to remove two screws on the side of the chassis to lift up the chassis a little bit so the cassette housing could sit properly. Once we got this housing into place, we screwed it down and placed a tape inside. Now when we pressed PLAY, we could see numbers changing on the display. All the functions were working properly again.

The sensor we changed is called a photointerrupter (Mitsubishi part number 268PO44010). It consists of an infrared emitting diode and phototransistor in a single package. The light from the diode shines up to the bottom of the reel. When the black surface is overhead, no light reflects to the phototransistor and output voltage of the photointerrupter is zero. When the shiny reflective surface is overhead, light reflects off the surface to the phototransistor and the output jumps to 1.2 V. As the reel turns, a squarewave is produced at the output. This is fed to the microprocessor, which contains a digital counter. Counter results appear on the VCR's display.

Some VCRs use one sensor, but this machine uses two, one for each reel. This adds a measure of security just in case the tape breaks and only one reel is spinning. The machine checks both reels and will shut down if only one is working. Figures G.1 to G.9 show how the repair was done.

Figure G-1.
The Mitsubishi HS-U59 VCR; notice the jog shuttle control on the right.

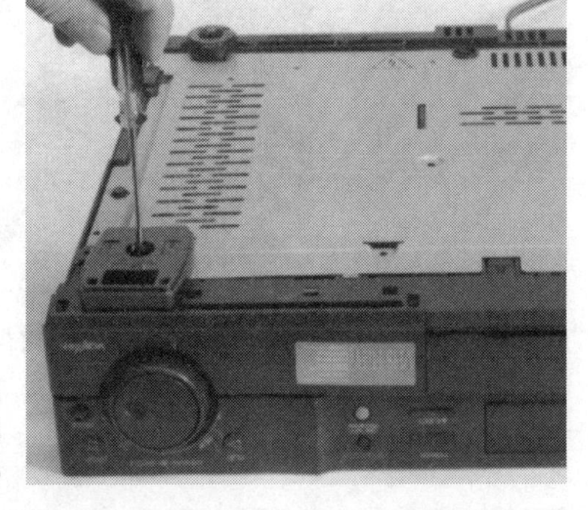

Figure G-2.
To remove the bottom cover, you have to remove four plastic feet.

Figure G-3. Lifting up the bottom cover of the VCR to get to the sensor's leads.

Figure G-4. Desoldering the sensor.

Figure G-5.
After detaching the front panel, removing the cassette housing.

Figure G-6.
Removing the cassette housing.

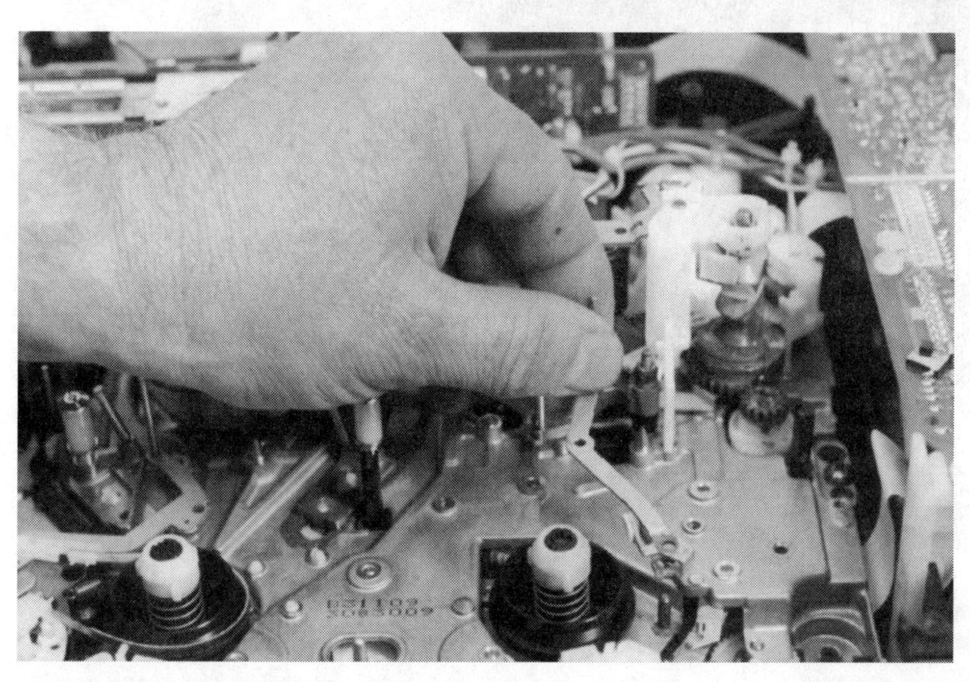

Figure G-7.
Removing a bracket.

Figure G-8. The pie shaped black-shiny pattern on the bottom of the take-up reel.

Figure G-9. Removing the defective sensor. Installing an exact replacement for this sensor solved the problem.

Appendix H

Case Study 8: RCA Model VR506A

This VCR had an intermittent clock display. In other words, the power would come and go. The moment you tried to activate any function, such as loading the tape, the power would die altogether. These symptoms are indicative of a problem in the power supply.

This VCR has a switch mode power supply. After removing the top cover of the VCR, we removed two screws from the power supply, one from the bottom of the VCR and one from the top, which attaches the supply to the chassis.

We serviced the supply as a separate unit. This is possible because the supply does not need a load to operate. When we measured the voltages at the secondary of the transformer with a DMM, they jumped around wildly from 0 to about 14 V. This measurement is taken at pin 5 the of output connector.

We checked three transistors to see if they were okay. We checked them in-circuit with the power off. We used the ohmmeter function of the DMM. All the transistors were good, no shorts or opens. When you check the resistance of a transistor between base and emitter, you get about 600 ohms if it is good. When you measure from emitter to base, you get infinite resistance. Between collector and base you get the same thing. Measuring between the collector and emitter doesn't normally show anything either way unless there is a short.

Next, we checked all the diodes. These also show about 600 ohms one way and infinite resistance the other way if they are good. These measurements made sense. Basically, these components should all be working if the power supply is coming on even for a short time. The fuse was okay, too.

The next obvious place to check is the feedback circuits of the switch mode supply. One obvious trouble spot is the feedback from pin 11 of the primary of the transformer PT1 to the base of the transistor Q1 through capacitor C09. This is a non-polarized capacitor. It is non-polarized because there is no DC voltage, only AC. This electrolytic capacitor looks just like a polarized capacitor, but it has no markings for plus and minus. In a pinch, you can put a polarized capacitor in this spot just to check if this is the problem. But, if you want the repair to be successful over the long term, you need to install an exact replacement part.

After we replaced this capacitor, we measured the voltages and they were rock solid. We put the supply back into place in the VCR. You don't have to worry about how to align the small connector on the supply (CN1) with the connector on the main circuit board (CN101). These go together quite easily as long as you align the two holes in the cover of the supply with the two plastic pins on the cabinet of the VCR. These, in effect, are your alignment guides. Once we maneuvered the power supply into place, we powered up the machine. Now all the functions worked normally. Figures H.1 to H.10 show how the repair was done.

Figure H-1.
The RCA VR506A VCR.

Figure H-2. To remove the power supply, you have to unscrew one screw from the bottom of the VCR.

Figure H-3. Removing a screw from the chassis.

Figure H-4. Lifting out the switch-mode power supply.

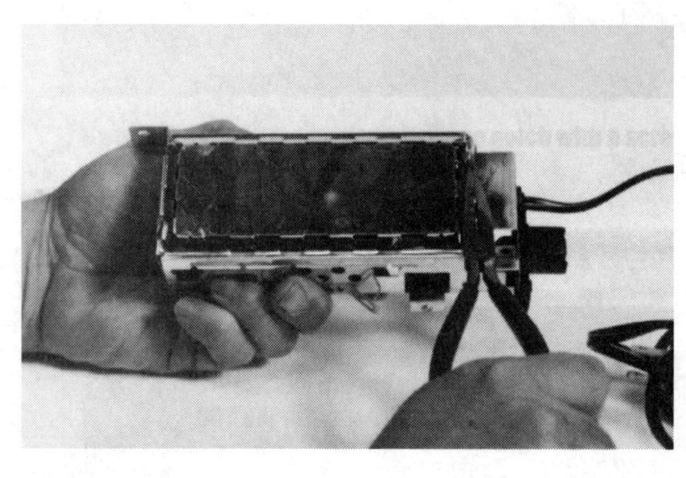

Figure H-5. Using the needle-nose pliers to straighten the tabs on the metal shield prior to removal.

Figure H-6. The foil side of the power supply.

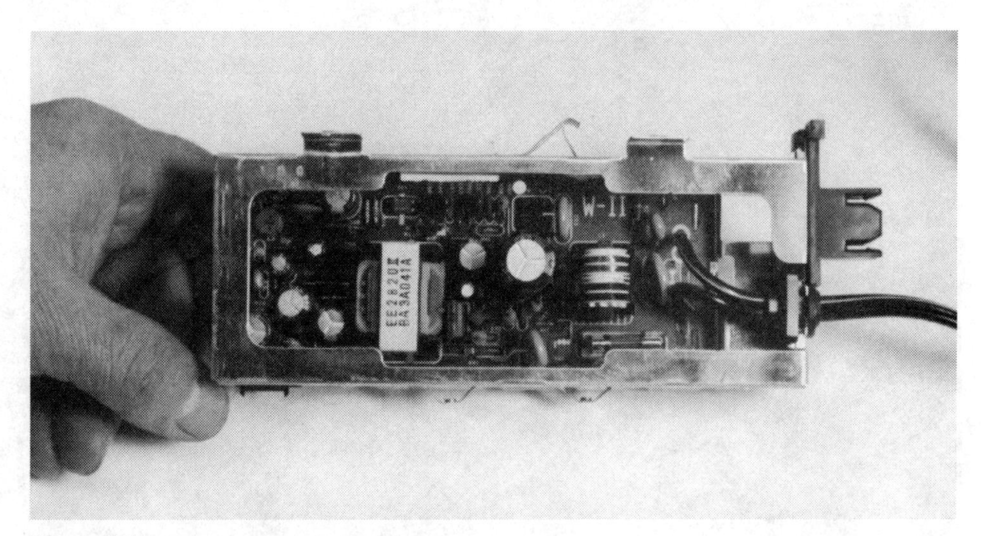

Figure H-7. The component side of the power supply.

Figure H-8. The defective capacitor in the power supply.

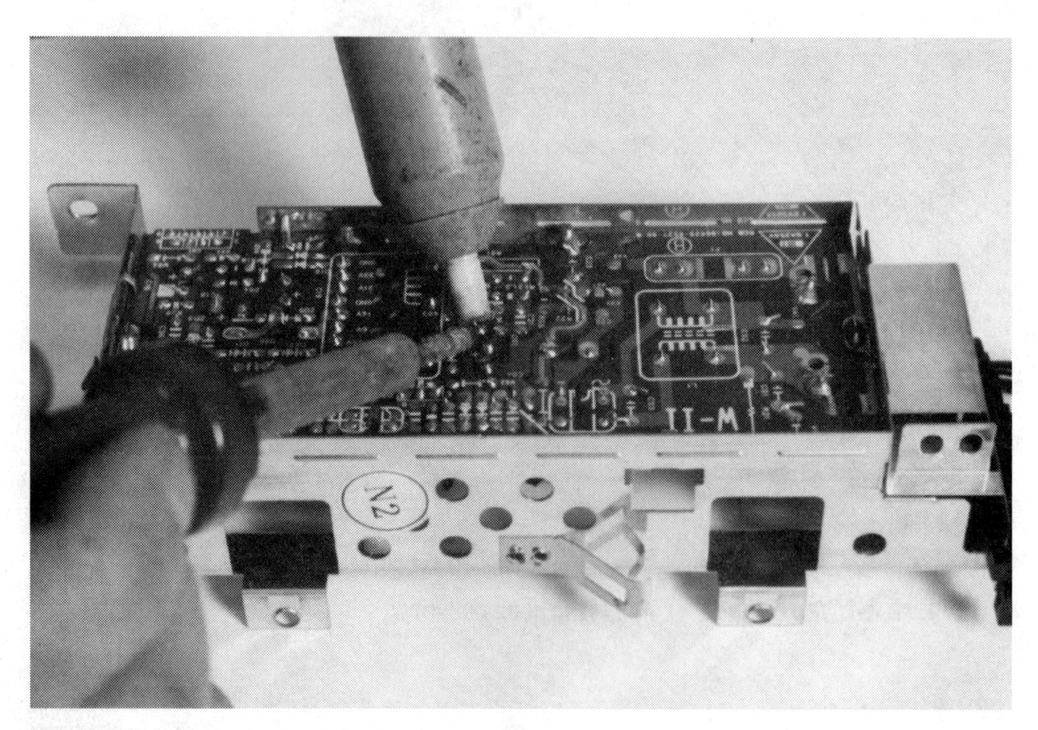

Figure H-9. Desoldering the defective capacitor.

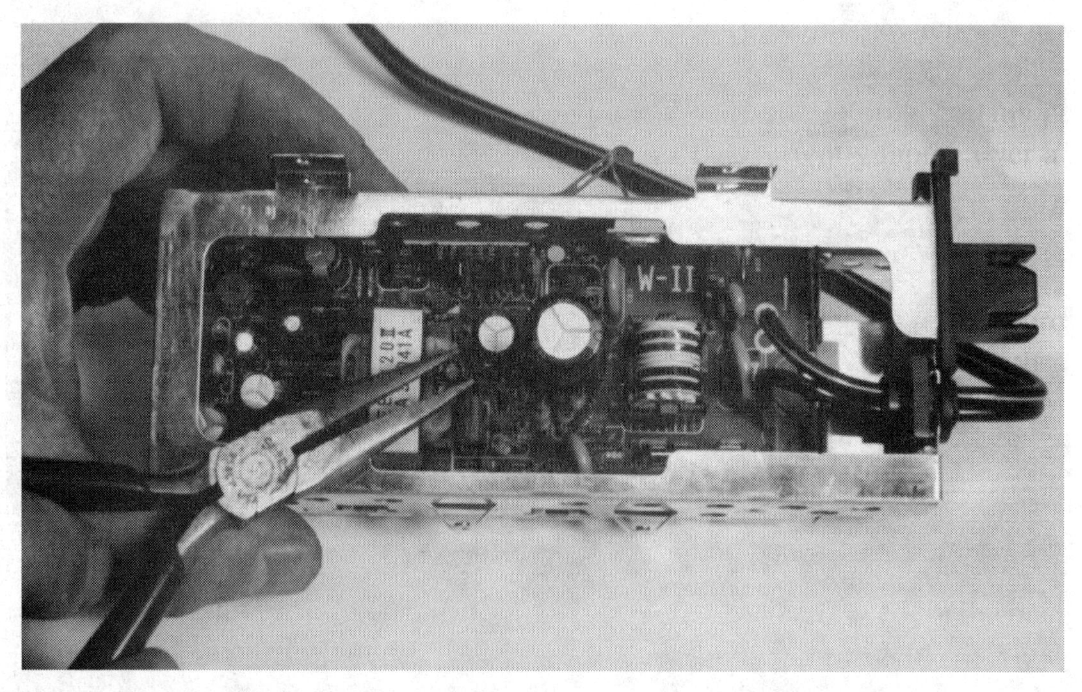

Figure H-10. Removing the defective capacitor with the pliers. We replaced this capacitor with another non-polarized capacitor of the same rating, which solved the problem.

Appendix I

Case Study 9: Sylvania Model VC4243AT01

This VCR would not rewind or fast forward. Sometimes this symptom indicates a worn out tire on the idler arm assembly. But, in this case, it was something altogether different.

This VCR utilizes a rubber bumper. This is a small piece of rubber that fits over an upright metal tab located right underneath the cassette housing on the right front side of the chassis. It is very easy to distinguish this rubber bumper, since it stays upright on the small metal tab. Next to the bumper is a white plastic bracket with a spring. It rests on the bumper when the VCR is in STOP mode.

When the VCR is engaged in fast forward or rewind, the plastic bracket is moved back and forth by a small toothed gear. In order for the plastic bracket to engage this gear, the bumper has to have sufficient thickness. If the rubber bumper is worn out, the bracket moves further away from the toothed gear. When fast forward is pressed, the toothed gear does not engage the plastic bracket.

This bracket is connected to the idler wheel. If the bracket doesn't move the idler wheel doesn't move and fast forward and rewind do not work.

It's difficult to gain access to the bumper. First, we removed the bottom cover. Next, we removed the belt that goes from the loading motor to the cassette housing pulley. Then, we removed the top cover and detached the front panel by releasing the plastic clips. There are six of them, three on the top and three on the bottom.

Next, we removed three screws on each side of the cassette housing, one large sheet metal screw in the middle and two smaller ones on either side of it. Then, we detached the plug on the right side of the assembly by pulling it to the right. Once we took all the screws out, we had to put one hand under the chassis and lift the front end of it up about an inch. While holding the chassis up, we gently removed the cassette housing by first lifting up the front part of the housing (because there are hinges in the back). The most difficult part of this operation was guiding the pulley on the cassette housing out of a tight spot between the chassis and the VCR cabinet.

Replacing the bumper was simple. Normally, you would slide the bumper up off the metal tab. In this case, though, the bumper was broken, so we just peeled it off the metal. You can purchase rubber bumpers in a package of ten from parts suppliers. If your regular supplier doesn't carry this part, you can order it from MAT Electronics (1-800-628-1118). We placed a new bumper on the metal tab, making sure the to push it to the bottom of the tab.

Reassembling the VCR is no less a feat than taking it apart. We inserted the back side of the cassette housing into the hinges and then again had to lift up the chassis with one hand and guide the worm gear pulley back into place. Then, we released the chassis and shook the cassette housing slightly to make sure it stayed in place. When it seemed to be seated properly, we put the screws back and continued the reassembly procedure making sure to reconnect the plug on the right side of the housing and reattach the belt that goes to the capstan motor. Figures I.1 to I.10 show how the repair was done.

Another VCR model with the same chassis (and thus employing the rubber bumper) is the Teknika VCR886. Both machines are made in Japan by Funai. If you ever have a question as to who manufactures a particular VCR, you can find out by looking at the FCC ID number at the back of the VCR. The Sylvania FCC ID is shown in Figure I.10. The FCC maintains databases of information on manufacturers who have applied for FCC certification. These databases are available for download on the Internet at www.fcc.gov. The FCC also maintains a bulletin board for information, which can be accessed by computer at 301-725-1072. Finally, the phone number of the FCC is 301-725-1585.

Figure I-1. The Sylvania VC4243AT01.

Figure I-2. Lifting up the bottom cover to examine the mechanism.

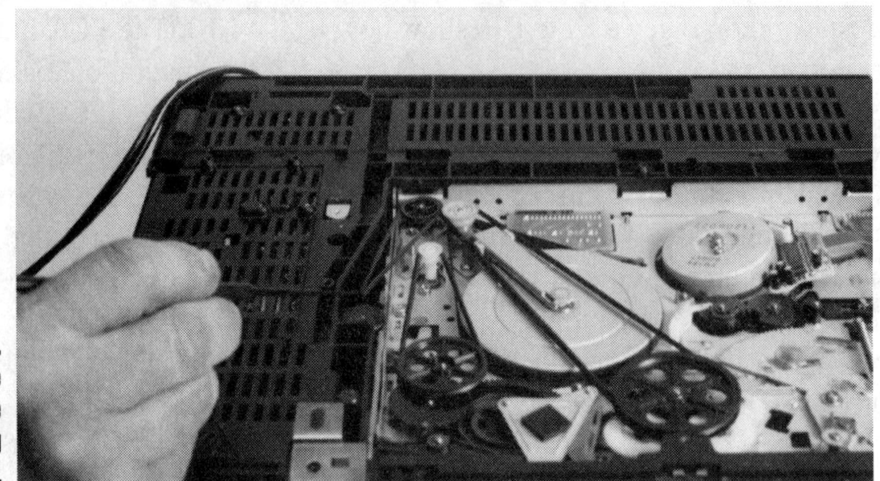

Figure I-3.
In order to remove the cassette housing from the VCR, we removed the loading belt with a special hook tool.

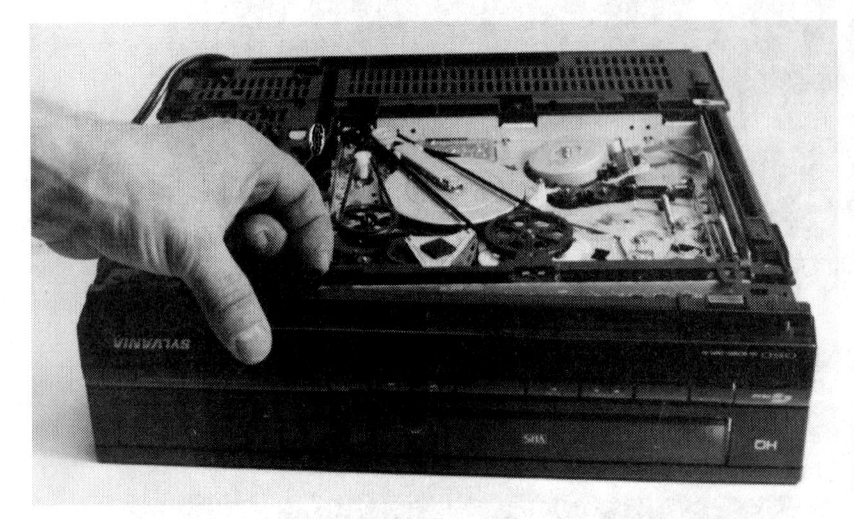

Figure I-4.
Unclipping the front panel.

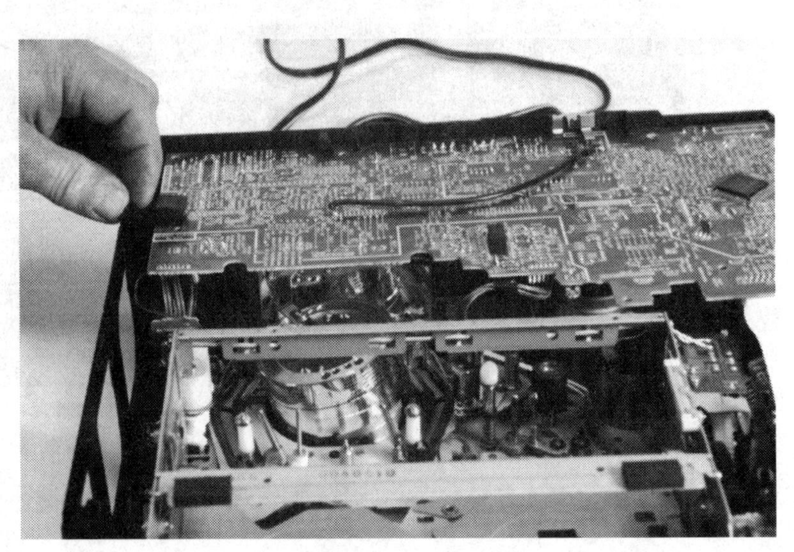

Figure I-5. Unclipping the main PC oard.

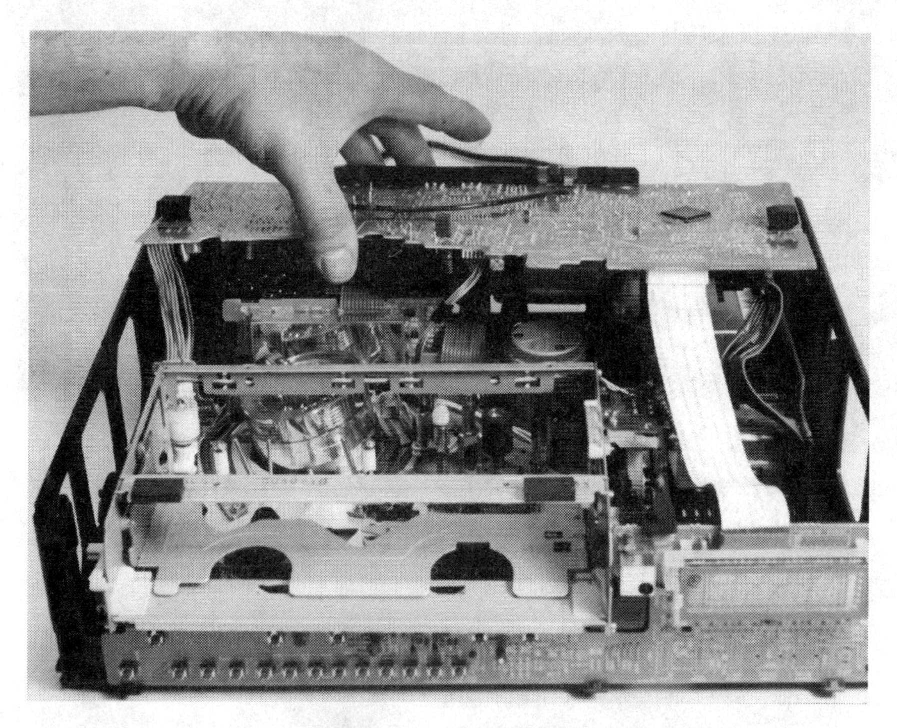

Figure I-6. Lifting the main PC board up and out of the way.

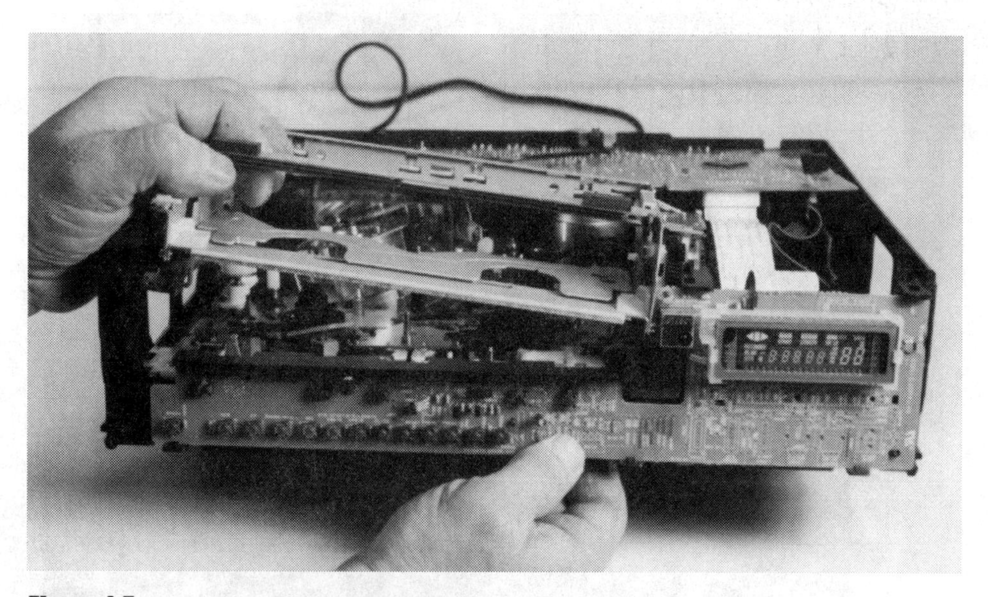

Figure I-7. Removing the cassette housing.

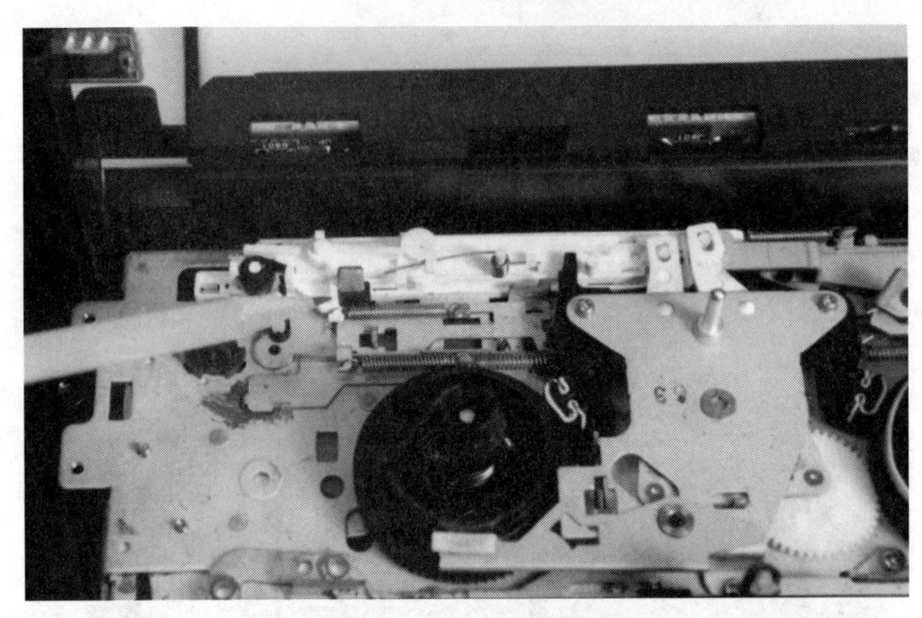

Figure I-8. A view of the rubber bumper on its metal tab from the rear of the VCR.

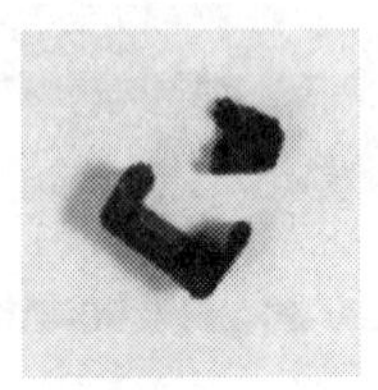

Figure I-9. A closeup of the rubber bumper, which came apart when we removed it.

Figure I-10. At the back of every VCR is a label with information, including the FCC ID. We found out that this VCR is manufactured by Funai.

Index

U

V

W

X